John Rickaby

**The First Principles of Knowledge**

John Rickaby

**The First Principles of Knowledge**

ISBN/EAN: 9783337248352

Printed in Europe, USA, Canada, Australia, Japan

Cover: Foto ©berggeist007 / pixelio.de

More available books at **www.hansebooks.com**

MANUALS OF CATHOLIC PHILOSOPHY.

ROEHAMPTON:
PRINTED BY JAMES STANLEY.

*MANUALS OF CATHOLIC PHILOSOPHY.*

# THE FIRST PRINCIPLES

OF

# KNOWLEDGE.

BY

JOHN RICKABY, S.J.

LONDON:
LONGMANS, GREEN & CO.
1888.

# PREFACE.

A FEW words will be enough to put exactly before the reader the object at which the present volume aims. A well-known criticism on the Aristotelian Logic is the complaint, that it provides for the consistency of thought with thought, but not for the consistency of thought with things; that it secures right processes upon given or assumed materials, but does not guarantee the materials upon which the processes are conducted. To supply the want thus indicated, several modern logicians have curtailed or omitted portions of the old Logic, and added new chapters, of which the following headings may serve as specimens, taken from Mr. Bain's work: " Uniformity and Laws of Nature," " Elimination of Cause and Effect," " Experimental Methods," "Frustration of the Methods," " Chance and its Eliminations," " Secondary Laws, Empirical and Derivative," " Explanation of Nature," " Hypotheses," " Classification," " Logic

*

of Mathematics," "Logic of Physics," "Logic of Chemistry," "Logic of Biology," "Logic of Rhetoric," "Logic of Politics," "Logic of Medicine." These titles show the kind of addition that now-a-days is asked, beyond the simple bill of fare found in the Aldrich who satisfied the students of a past generation, and to many even afforded more than they wanted.

It is unfortunate that those who in this country were, perhaps, the loudest in their clamours that logic should take account of the reality which hitherto it had seemed to neglect, should have embraced a system of philosophy which is fatal to firm belief in any reality beyond thought itself. Messrs. Mill and Bain assuredly have not directly tended to take men out of idealism, and make them realists. Yet the former was explicit enough in his demands:[1] "I conceive it to be true, that Logic is not the theory of thought as thought, but of valid thought: not of thinking, but of correct thinking.... In no case can the thinking be valid unless the concepts, judgments, and conclusions resulting from it are conformable to fact. And in no case can we satisfy ourselves that they are so by looking merely at the relations of one part of the train of thought with another. We must ascend to the original

[1] *Examination*, c. xx. pp. 397, seq. (2nd Edit.)

sources, the presentations of experience, and examine the train of thought in relation to these."

Little as the modern representatives of the Schoolmen are satisfied, either with the spirit of Mr. Mill's demand, or with the mode of his own response to it, they have deemed it well worth while, not indeed to change the old Logic, but to add to it a new book. Pure Logic remains substantially what it was, and is justified in its position. It assumes, as all other sciences do and must, that human thought has, in general, objective reality; and on this most legitimate assumption it proceeds to lay down the laws of orderly, consistent thinking. The newly added part of Logic, often called Material, Applied or Critical, takes for its special purpose to defend the objective reality of thought. It is thus an assertion of a form of realism, as against idealism, and is called in this book the *Philosophy of Certitude*. For the whole question comes to this: what reasonable account can be given of man's claim to have real certainty about things? What are the ultimate grounds for holding, that man may regard his knowledge about objects as undoubtedly correct? Scientifically to draw out the account here demanded is a work appositely described by the title, *The First Principles of Knowledge*.

An endeavour has been made throughout these pages, while stating the sound, traditional principles of certitude, to bring them into constant contact with the antagonist principles, more particularly with the principles of Hume and the pure empirics. It is not true that the only possible philosophy is a history of the opinions which, at various times, have prevailed; but it is true, that the modern spirit will not be satisfied without a statement of how controversies stand on questions which are notoriously disputed. The truth as made manifest in conflict, is what has to be exhibited : and this necessity, whether exactly desirable or not, must stand as explanation or apology to those, whose own special tastes might prompt them to desire a simple exposition of scholastic doctrine apart from the encumbrance of adverse systems. Scholasticism must now be militant, and that, not only with a view to outsiders, but with a view to retaining its own clients, who cannot fail to come across much in modern literature, for the understanding and the consequent rejection of which some direct preparation is needful.

Readers not already familiar with the questions here discussed, would do well at first to leave alone the notes which are printed in smaller type, and concentrate attention on the positive doctrine, the importance of which must be judged, not by the length of its statement, but by the weight of the words. The matter is eminently one which is best conveyed in a few precise sentences, the full import of which must be mastered by leisurely consideration.

# CONTENTS.

## PART I.—THE NATURE OF CERTITUDE IN GENERAL.

|   | PAGE |
|---|---|
| CHAPTER I.—DEFINITION OF TRUTH | 1 |
| ,, II.—IN WHAT ACT OF THE MIND A TRUTH MAY BE FOUND COMPLETELY POSSESSED | 14 |
| ,, III.—DEFINITION OF CERTITUDE AND OF THE STATES OF MIND FALLING SHORT OF CERTITUDE | 42 |
| ,, IV.—KINDS AND DEGREES OF CERTITUDE | 50 |
| ,, V.—METAPHYSICAL AND PHYSICAL CERTITUDE | 68 |
| ,, VI.—THE ORDER OF PRECEDENCE BETWEEN NATURAL AND PHILOSOPHIC CERTITUDE | 108 |
| ,, VII.—THE CHARGE OF DISCORD (OR AT LEAST OF WANT OF CO-OPERATION) BETWEEN NATURAL AND PHILOSOPHIC CERTITUDE | 119 |
| ,, VIII.—UNIVERSAL SCEPTICISM | 134 |
| ,, IX.—CARTESIAN DOUBT | 148 |
| ,, X.—THE PRIMARY FACTS AND PRINCIPLES OF THE LOGICIAN | 164 |
| ,, XI.—RETROSPECT AND PROSPECT | 183 |
| ,, XII.—THE REJECTION OF VARIOUS THEORIES ABOUT THE ULTIMATE CRITERION OF CERTITUDE | 188 |
| ,, XIII.—EVIDENCE AS THE ULTIMATE OBJECTIVE CRITERION OF TRUTH | 216 |
| ,, XIV.—THE ORIGIN OF ERROR IN THE UNDERSTANDING | 232 |

## PART II.—SPECIAL TREATMENT OF CERTITUDE.

|  | PAGE |
|---|---|
| CHAPTER I.—SHORT INTRODUCTION | 249 |
| ,, II.—THE TRUSTWORTHINESS OF THE SENSES | 258 |
| ,, III.—OBJECTIVITY OF IDEAS, WHETHER SINGULAR OR UNIVERSAL | 301 |
| ,, IV.—EXAGGERATED REALISM, NOMINALISM, AND CONCEPTUALISM | 332 |
| ,, V.—CONSCIOUSNESS | 340 |
| ,, VI.—MEMORY | 366 |
| ,, VII.—BELIEF ON HUMAN TESTIMONY | 377 |
| ,, VIII.—BELIEF ON DIVINE TESTIMONY | 389 |

# THE FIRST PRINCIPLES OF KNOWLEDGE.

## PART I.
## THE NATURE OF CERTITUDE IN GENERAL.

### CHAPTER I.
#### DEFINITION OF TRUTH.

*Synopsis.*
1. Three kinds of truth.
2. Definition of truth as found in knowledge. (*a*) The common-sense view, agreeing with the scholastic definition. (*b*) Assigned reasons for modifying and even radically altering that definition. (*c*) Assertion of the old definition, a course which the rest of this work must defend, but for which a little may be said at the outset.
3. Definition of error, as a corollary to the definition of truth.

*Addenda.*

1. TRUTH is commonly divided into truth of things, truth of thought about things, and truth in the outward expression of our thought about things. The first kind of truth is called ontological, the third moral, and each of these is discussed in separate volumes of the present series. It is with the second member of this division, about what is often styled logical truth, that the treatise which we are here beginning is concerned. What true knowledge is, and how its possession by the human

intellect can be vindicated, these are the questions specially calling for our investigation.

2. (*a*) An ordinary man, if asked to explain what true knowledge is, would reply simply, that knowledge was true when the thing was really such as we thought it to be. He would thus agree with the definition of the schoolmen, "Truth is an equation or a conformity of thought to thing."[1]

(*b*) But a matter which, at first sight, seems thus readily settled, presents, on reflexion, a number of difficulties, which some have regarded as so serious as to upset the plain man's view, backed though it be, by other philosophies, and by the massive volumes of scholastic philosophy which the centuries have piled one on the top of another. For when the case is more narrowly sifted, are we not driven to make some awkward inquiries: How can mental images be like outer objects, especially material objects? How are sensations like the external bodies which stimulate them? When several senses give their reports of one object, as there is no likeness between the several reports, say between the taste of an orange and its colour, how can they be all like the object? What is the conclusion from the notorious fact that in different persons, and even in the same person at different times, any one of the five senses may bear divergent testimonies? Besides, even if our knowledge were like what we call its object, to observe this correspondence must be an act of comparison, which not we, but some one else, must make: for we at least

[1] Tongiorgi, *Institutiones Philosophicæ*, Vol. I. nn. 370, seq.

cannot compare the thing as known to us, with the thing as out of our knowledge. Such an attempt on our part would be preposterous, a fraudulent endeavour to assert, for our essentially relative faculties, an absolute validity.

Moved by these considerations, a number of modern philosophers dare to claim for human knowledge only some correspondence with its object which is less than that of likeness, and is describable as symbolic: just as a mathematical formula, though not like, may yet symbolize, the path of a cannon-ball. Thus, in the words of Mr. Frederick Harrison, "our scientific conceptions have a very good working correspondence with the assumed reality without; but we have no means of knowing whether the absolute correspondence between them be great or small, or whether there be any absolute correspondence at all." Mr. H. Spencer,[2] in expounding his theory of "transfigured realism," sets forth still more clearly the position of the "relativist," that is, of the philosopher who denies that we can know things absolutely as they are. Maintaining that "*resistance*, as disclosed by opposition to our energies, is the only species of external activity which we are obliged to think of as subjectively and objectively the same," still even here he will not positively affirm that knowledge is like the object. And for ordinary objects his teaching is this: "If $x$ and $y$ are the two uniformly connected properties in some

[2] Mr. Spencer's doctrine may be seen in his *Psychology*, Part I. c. xix.; *First Principles*, Part I. c. ii.

outer object, while *a* and *b* are the effects which they produce in our consciousness; the sole need is that *a* and *b*, and the relation between them, shall always answer to $x$ and $y$, and the relation between them. It matters not if *a* and *b* are both like $x$ and $y$, or not; could they be exactly identical with them we should not be one whit the better, and their total dissimilarity is no disadvantage."[3] In other words, if for every definite change in objects, there is one constant change in the mind which is affected by that object, then this is enough, without any resemblance of thought to thing; concomitant variation suffices.

(*c*) Against this theory the time-honoured definition of knowledge must be re-affirmed. The objections raised against it are only the old arguments in favour of complete distrust in the power of man to attain truth; and to refute them will be the main purpose of all that follows in this volume. At the outset, this book defines truth of intellect to be "the conformity of thought to thing": subsequently its one grand aim will be gradually to make good the definition. Whilst patiently awaiting the development of a long line of argument, the reader may find some consolation in a few declarations that can be offered him at once. First of all, when knowledge is said to be a sort of "equation" of mind to thing, it is not meant that knowledge, in order to be true, must exhaust the whole object: a partial knowledge is true, as far as it goes, especially when it is recognized as only partial.

[3] *First Principles*, Part I. c. iv. § 25.

## DEFINITION OF TRUTH.

Next, the likeness which is asserted is quite *sui generis*. An idea,—to use the word at present in its broad sense of any act of knowledge,—is not a dead picture, but something effected by and in the living, cognitive mind; it is a thing with a conscious meaning of its own:[4] it is, as Spinoza says, self-assertive or self-referent; it is what the schoolmen sometimes call a *signum quo*, a sign which taken, not in its isolation as a mere phenomenon, but as it exists in the mind, is the knowledge of the thing signified. Thus it differs from a *signum ex quo*, a sign which has first to be known, that from it the mind may travel to the thing signified. To quote Father Liberatore: "The *signum a quo* is that which, by being first known, leads on the mind to the knowledge of the thing it signifies. Another way of signifying is presented to us in those inward signs which do not come before the mind as objects of its perceptions, but which, by informing the cognitive faculty, effect actual knowledge. These latter may be called *signa in quibus* [or, *signa quibus*], or also, 'formal signs,' in that they do not represent objects as previously known, but are forms determining the mind to perceive the object. To this category belong mental concepts."[5] Hence the representative, significative, or meaning power of an idea is not photographic, nor anything analogous to photography; and to fancy it so is the cardinal

---

[4] Hence Hume greatly errs: "The reference of an idea to an object is an extraneous denomination, of which in itself it bears no mark or character." (*Treatise*, Part I. sec. vii. pp. 327, 330.)

[5] " Signum *ex quo* quod prius in se cognoscatur et ex sui cognitione potentiam ducat in cognitionem rei significatæ. Alter modus significandi locum habet in signis internis, quæ non se offerunt ut objecta,

error of scepticism. The idea depicts its object in no way open to our artists: neither by similarity of substance, nor of colour, nor of outline, nor by any mode of material portraiture. The process is so peculiarly mental or spiritual, that illustrations borrowed from matter are more calculated to mislead than to direct. The uniqueness of the phenomenon is essentially its strangeness, for we cannot explain it by reduction to any familiar class. Yet the strangeness is welcome as serving, in another treatise, to show the inadequacy of the materialistic hypothesis.

A difficulty, raised by Cousin, really amounts to no more than a matter of words. He says[6] that an idea cannot be "an image" of an object, because only material representations can be "images;" that we cannot strictly speak of "likeness" between spiritual objects, but only between material: and that therefore, if we do call ideas "images" bearing the "likeness" of their objects, we are talking not properly but metaphorically. As our knowledge begins in sense, so far all our spiritual ideas may be said to be conveyed in metaphorical terms; it seems however a fair procedure to regard a term as no longer metaphorical, when we no longer advert to the figurative meaning, but pass straight to the main object. Thus, in speaking of moral rectitude

sed informando potentiam eam efficiunt cognoscentem in actu. Hæc dici possunt signa *in quibus* [vel signa *quibus*], vel etiam signa *formalia :* quia non repræsentant tanquam objecta prius cognita sed tanquam formæ determinantes potentiam ad perceptionem objecti. Hujusmodi est conceptus mentis." (*Logica*, Pars I. c. i. n. 5.)

[6] *Histoire de la Philosophie*, leçon 21me, et alibi passim.

we hardly refer to the image of a man keeping the straight path as he walks; but we go at once to the notion of right conduct. Similarly, whatever may have been the origin of the terms, we can now claim to apply the word "image" or "likeness" straight to spiritual resemblances. However, if any one should insist on seeing a trace of metaphor left here, the point is not worth controverting.

We assume, therefore, that ideas are, in the language of Aristotle, ὁμοιώματα τῶν πραγμάτων, —"likenesses of things," and so stand contrasted with words which are conventional signs: whereby a special meaning is given to the saying, that man is a microcosm.[7] Not only does man sum up the several constituents of our Cosmos by uniting together mineral, vegetable, and animal nature; but by knowing all things, he, in a manner, reproduces, or becomes all things. *Homo est quod est*, says the materialist scoffingly; translating the words, "Man is what he eats, what his food makes him." False as a complete statement, this is true as a partial statement; and equally true is it, *homo est quod scit*—"Man is what he knows." In this sense St. Thomas writes *anima est quodammodo omnia*— "The soul is in some sense everything;" which is the repetition of Aristotle,[8] ἡ ψυχὴ τὰ ὄντα πώς ἐστι πάντα—"The soul is in a manner all things."

3. From the definition of true knowledge before given, a corollary as to the nature of error may be gathered. Not any absence of likeness between thought and thing is straightway falsehood: rather

---

7 Silvester Maurus, *Quæstiones Philosophicæ*, quæstio vi.
8 *De Anima*, Lib. III. c. viii. 1.

such mere absence is ignorance. Before downright error is reached, there must be not only want of conformity but positive deformity. For knowledge, however limited, is true knowledge so long as it does not transgress or deny its own limits—a fact highly important to finite intelligences like ourselves who can but "know in part." Be it our consolation, then, to remember that the opposition between knowledge and ignorance is only what the logicians call a "contradictory;" while it is not till we have gone as far as the "contrary" opposition that we commit error.

### Addenda.

(1) The mysteriousness of the act of knowledge, and its apparent impossibility on any material analogy, were points that arrested the attention of the early speculators. Whereas St. Thomas[1] argued that because the mind was capable of becoming cognizant of bodies it could not itself be corporeal, some of the old Greeks had pushed to its extremes the principle, Like is known by like — ὅμοια ὁμοίοις γιγνώσκεται. Empedocles, for example, had said, "We perceive earth by means of earth, water by means of water, air by means of air, fire by means of fire, love by means of love, and strife by means of strife," where love and strife stand for what we call attractive and repulsive forces. Others spoke either of the eye sending out its influences to the object, or of the object emitting its εἴδωλα, or minute images to the eye, which,

> Like little films from outer surface torn,
> In mid air, to and fro, are lightly borne.[2]

[1] 1a, q. 75. art. 2.
[2] Quæ quasi membranæ, summo de corpore rerum
Dereptæ, volitant ultroque citroque per auras.
(*Lucretius*, iv. 31, 32.)

All such conceptions are the follies of a crude materialism; and a long way the better course is, while admitting that *how* knowledge is possible is inscrutable to us, yet to insist that the fact is manifest to experience. The reaction of our faculties to their appropriate *stimuli* must simply be accepted for what it declares itself to be, namely, not any kind of a reaction, but the special reaction which must be called cognitive.

(2) Preferring theory to fact, and a theory the very arguments for which rest on an assumption which is just the contrary to what they are fancied to prove, a number of modern writers quite set aside the doctrine that we have knowledge like to the realities outside of ourselves. From America comes the voice of Mr. Borden Browne, telling us, as an introduction to his volume on *Metaphysics*, that because we cannot compare thought with being, therefore, "truth cannot be viewed as the correspondence of thought and thing, but as the universally valid in our thought of the thing. That is the true conception of reality, which grasps the common to all, and not the special to one;" so that the test of truth is "the necessity of the conception and the inner harmony between several conceptions. It is not the lack of harmony between our conceptions and reality which disturbs, but the discord of our conceptions among themselves." A like utterance we have from a German author, according to whom "truth does not consist in any sort of correspondence between our thought and the things outside us, but in a character that belongs to our mode of putting together our internal experiences. Our thoughts are true, when their nature, as internal events, is understood; when they are placed in equal relation to the rest of experience. The criterion of truth is the feeling of universality and necessity in the ultimate axioms." In our own country we have

some authors completely rejecting the doctrine that truth is conformity of mind to thing; while others, using the same words, are less thorough in their divergence from us, though sufficiently divergent to be in decided opposition. They distinguish between perception and thought, so as to make thought more especially a matter of subjective forms without ascertainable objective validity. "Truth relatively to the human mind," writes Mansel,[3] "cannot be defined as a conformity with its object; for to us the object exists only as it is known by one or other faculty. Hence *material* truth consists rather in the conformity of the object as represented in thought with the object as presented in intuition; while logical truth consists in the conformity of thought to its own laws." With these words may be compared the following from another countryman of ours, Mr. S. Hodgson: "Without thought no truth, without perception no reality. By reality I understand the actual existence of the object, its actual presence to consciousness.[4] Reality is not greater after thought than before; thought has transformed it into a new shape, has given it new relations, but has added nothing to its real existence. Truth, on the other hand, is a product of thought, the form which an object assumes after investigation, and is thus greater after thought than before it. Reality depends on the relation between objects and consciousness: truth on the relation between objects in consciousness." The difference here is partly a matter of words, but it is also a matter of fundamental doctrine; and with regard to each of the authors cited, it is sufficient for present purposes if the reader

[3] *Prolegomena Logica*, c. vi. in fine; also Mansel's *Aldrich*, Appendix M. p. 277. (3rd Ed.)

[4] See Green on *Thought as constitutive of the Reality of the World; Introduction to Hume*, § 173. The passage from Mr. Hodgson is found in his book on *Time and Space*, p. 352.

understands them as representatives of a now widespread revolt against the scholastic definition, "All truth in cognition consists of the assimilation or conformity of the mind with the object."[5]

(3) The consequences of the doctrine logically carried out, that ideas are mere symbols, are very fatal to all religious belief, as well as to everything worth calling true knowledge: and though it is often protested, that a doctrine is not to be judged by its inconvenient logical consequences, but by its intrinsic truth, yet there are consequences so manifestly bad, as to afford evidence that the premisses, whence they are drawn, cannot be sound. If our knowledge of things is what adversaries say it is, then it is not genuine knowledge at all: and this some of themselves admit. Take, for instance, Lange's confession, in words gathered from a long declaration: "All our knowledge of nature is, in fact, no knowledge at all, and affords us merely the substitute for an explanation. The intelligible world is a world of poesy, and precisely upon this fact rests its worth and nobility. No thought is so calculated to reconcile poesy and science as the thought, that all our reality is only appearance."[6] So, too, Mr. Spencer[7] is perpetually harping on the string, "ultimate religious ideas, and ultimate scientific ideas are mere symbols of the actual, not cognitions;" even "the personality of which each is conscious, and of which the existence is to each a fact beyond all others the most certain, cannot be truly known at all: knowledge of it is forbidden by the very nature of thought." It is well that the reader should thus be brought

[5] "Veritas in cognoscendo est mentis assimilatio vel conformitas ad rem."

[6] *History of Materialism* (English Translation), Vol. II. p. 309.

[7] See the conclusions to the early part of *First Principles*, Part I. cc. iv. and v., Part II. c. iii.

plainly to see into what a gulf of nescience he is about to plunge, if he is resolved to take up the theory, that the old definition of man's actual knowledge must be abandoned. Ultimately he must be driven to say with Fichte,[8] " Reality all merges into a marvellous dream, without life to dream about or spirit to dream—a dream which is gathered up into a dream of itself."

(4) In unconscious anticipation of modern difficulties, the scholastics strongly insisted on knowledge as being *mental assimilation;* in proof of which assertion we will borrow a few citations made by Kleutgen on this subject.[9] " Every cognition is brought about by the likeness of the object known, in the mind that knows."[10] "In the first place, we suppose it to be essential to the act of the intellect, aye, and to every cognition, that a certain assimilation be produced in the mind of the intelligent agent. This fundamental position may be taken to be a dogma and a principle both in theology and in philosophy, questioned by none."[11] " It can in no wise be denied, that when the rational mind, reflecting on itself, becomes self-conscious, a likeness of itself is produced by this cognition, or even that this cognition is an image of self, fashioned after its own likeness, as it were an impression of self upon self."[12] Theologically the doctrine is

[8] Quoted in *Hamilton's Reid*, p. 129, note.
[9] *Die Philosophie der Vorzeit*, I. i. §§ 23—25.
[10] " Omnis cognitio fit secundum similitudinem cogniti in cognoscente." (St. Thomas, *Contra Gentes*, Lib. ii. c. 77.)
[11] " In primis supponimus de ratione intellectionis, imo et cognitionis esse, ut per quamdam assimilationem intra mentem intelligentis fiat. Hoc fundamentum videtur esse veluti dogma, et principium in philosophia et theologia communi consensu receptum." (Suarez, *De Angelis*, Lib. ii. c. 3; *De Anima*, Lib. iii. c. i.)
[12] "Nulla ratione negari potest, cum mens rationalis se ipsam cogitando intelligit, imaginem ipsius nasci in sua cognitione; imo ipsam cognitionem sui esse suam imaginem ad sui similitudinem, tanquam ex ejus impressione formatam." (St. Anselm, *Monol.*, c. 33.)

## DEFINITION OF TRUTH.

of importance in reference to the Blessed Trinity, in which the Son is begotten in the likeness of the Father, as the Father's Word, or intelligible term.[13] Silvester Maurus has some apposite remarks: "The procession of the Son is of such sort as to express the Father in His nature and essence. The act of the intellect, whereby we know ourselves, is likewise posited for the purpose of expressing the intelligent agent in his essence and nature, into which the intellect alone can penetrate. Since in God, the Word, *i.e.*, the term produced by the act of the intellect, receives the self-same nature and essence with the Father, His production or procession is hence fitly termed a generation, that is, the origin of a living subject from a conjoint living principle, from whom it receives a similar nature."[14]

[13] Heb. i. 3; Coloss. i. 15.
[14] "Filius producitur ad hunc finem ut exprimat patrem in natura et essentia. Intellectio, qua quis intelligit seipsum, producitur in hunc finem ut exprimat intelligentem in natura et essentia, quam penetrat solus intellectus. In divinis intellectio producta, seu Verbum, accipit eandem numero naturam et essentiam Patris: ergo ejus productio proprie est generatio, hoc est, origo viventis a vivente principio conjuncto, in similitudinem naturæ." (*Quæstiones Philosophicæ*, q. ii.)

## CHAPTER II.

### IN WHAT ACT OF THE MIND A TRUTH MAY BE FOUND COMPLETELY POSSESSED.

*Synopsis.*
1. Division of the mind's acts into three, Apprehension, Judgment, Reasoning.
2. It is in the judgment that a truth may be found completely possessed.
3. Hereon certain discussions arise. (*a*) The various definitions of judgment. (*b*) Suggestions on the subject from comparative philology. (*c*) A view taken of judgment by St. Thomas.

*Addenda.*

1. FOR their own convenience logicians have long been accustomed to divide the acts of intellect into three. The mind in viewing an object may be regarded either as making an affirmation or a denial about it, or else as not affirming or denying. In the last case the act is called an *Apprehension;* in the first case it is called simply a *Judgment,* when the decision is immediate, and *Reasoning* or *Ratiocination,* when the decision is mediate, a conclusion drawn from previous judgments. Now the question to be raised is, to which of these acts does the complete grasp of a truth belong; and because between an immediate judgment and a mediate judgment the difference does not affect the present inquiry, the selection lies

between *Apprehension* and *Judgment*. Thus the threefold division is no longer necessary: a twofold suffices. Throughout, however, the reference is only to human modes of knowledge, not to those higher modes which transcend our comparatively imperfect act of judgment.

On the threshold, the investigation seems to be stopped by serious doubts which may be started, as to whether any act of apprehension is simply such, and not also a judgment—a judgment on some point, if not precisely on the definite point proposed. What is meant is, that when, for example, the proposition, "Quinine will benefit the patient," passes through a physician's mind, he may very well, for lack of evidence, leave the main judgment unformed; but all the same, some contend that he cannot, without forming upon them any judgment whatever, simply apprehend the two terms, "quinine" and "beneficial to the patient." Still less could he do this in an analytical proposition such as "The whole is greater than its part."

We need not decide this controversy at starting, but will return to it presently. Meantime we proceed thus. Instead of the definition, "Apprehension is the act of the mind *which* neither affirms nor denies," we have but to substitute, "Apprehension is the act of the mind *so far as* it neither affirms nor denies, but merely places an object before the consciousness." Then if the distinction between apprehension and judgment should prove to be, not real, but only the result of a mental abstraction, or only a difference between a judgment of one order and a judgment of another order; still it would be available for the discussion in which it is now to be used.

2. An apprehension, as above defined, cannot be false: what is apprehended is so far truly apprehended and cannot be otherwise. The object before the mind must, of course, be the object before the mind; just as what a man sees, with his eyes, that he sees, even though he should, by a mistake in inference, proceed to name it wrongly. But while apprehension enjoys this immunity from error, it has the countervailing disadvantage that it never fully contains a truth: and here is just the fact which has to be brought out. Unless the mind has equivalently got as far as an affirmation or a denial, it has not completely possessed itself of a truth. No man can claim the merit of having uttered great political truths, if he has only thrown out a number of terms as in apprehension, not as combined into judgments: "force of popular will," "resistance of the wiser few to the ignorant many;" "adaptability to circumstances," "fixity of principle;" "generous liberalism," "prudent conservatism," and so forth. These are terms which might occur in any one's speech, no matter what were his opinions. The case is so clear, that it is hardly needful to amplify the bare statement of it; though it may be useful to note that a student might confuse himself, if, without warning, he were to light on some very self-evident proposition and test the doctrine by it; thus, "The whole is greater than its part." But here, however inevitable and simultaneous the judg-

ment may be, it is not exactly identical with the apprehension as limited by the definition already given.

In every instance, then, to affirm or to deny, to say *is* or *is not*, this is the point where, and where alone, the mind fully commits itself to a truth or to a falsehood. To make some assertion, positive or negative, *there* lies the risk, *there* is the success or failure, so far as truth is concerned. A mere apprehension is a step along the right road, but it does not quite reach the goal. Hence the need of insisting on the judgment as the great crowning act in the order of intelligence; and of giving to "is" and "is not" a most prominent position in the science of logic. Grammarians may settle among themselves how much or how little they will put into their definition of a verb; but for logicians the words of St. Thomas must be the guide: "Intellectual truth consists in the equation between the mind and reality, in consequence of which the mind affirms that the object is that which it really is, or denies it to be what it really is not."[1]

3. Still it must not be pretended that in assigning to apprehension a definition, which shirked the real difficulty of its distinction from judgment, an author has fulfilled all justice. We must solve the doubts already suggested.

(*a*) There is an awkwardness, at the outset, about the definition of a judgment—what precisely it is. Of proposed definitions some obviously have

[1] "Veritas intellectus est adæquatio intellectus et rei, secundum quod intellectus dicit esse quod est, et non esse quod non est." (*Contra Gentes*, Lib. I. c. lix.)

C

no proper title to that name, being rather things that may be affirmed about propositions, than accounts of the very nature of propositions. In treating the subject with a view to definition, some authors prefer to represent the predicate as containing the subject, others the subject as containing the predicate; a difference that amounts to one which, in pure logic, is styled that between "extension" and "comprehension" or "intension." In the enunciation, "Man is an animal," "animal" may be regarded, in extension, as a class under which man is included; or "man" may be regarded, in comprehension, as including a number of constituent notes, of which animality is one. A third method is to put subject and object on a line of equality, instead of on a scale of subordination. Such is the tendency of the following definitions taken in order from Hobbes, James Mill, and John S. Mill.[2] "In every proposition the thing signified is the belief, that the predicate is the name of the same thing of which the subject is a name:" "Predication consists essentially in the application of two marks to the same thing:" "According to the formula best adapted to express the import of a proposition as a portion of our theoretical knowledge, *all men are mortal*, means, that the attributes of man are all accompanied by the attributes of mortality;" while from another point of view the best formula is, "The attributes of man

---

[2] See James Mill's *Analysis*, c. iv. s. iv.; John Stuart Mill's *Logic*, Vol. I. Bk. I. cc. v. and vi., where the quotation from Hobbes is given; see also *Leviathan*, Part I. c. iv. p. 23. (Molesworth's Edition.)

are evidence of, a mark of, mortality." The co-ordination of subject and predicate is still more secured by the device of Mr. F. H. Bradley,[3] who regards the simple judgment as containing, not two ideas, but one compound idea, which the judgment "refers off to the region of reality." Thus, *the wolf is eating the lamb* is interpreted as assigning over to reality the complex notion *wolf-eating-lamb:* wolf-eating-lamb is a reality or fact. This way of regarding the matter at least calls attention to an important truth in logic, namely, that judgment is not simply any mode of linking ideas together, even though there be no copula and nothing equivalent to it. Nor is Mr. Bradley's view to be confounded with the extravagant theory of Antisthenes and others,[4] to the effect that the only valid judgments are those in which subject and predicate are identical (Aristotle, *Metaph.*, v. 29): for he does not maintain, that in the proposition, "The wolf is eating the lamb," predicate and subject are one in the fullest sense of oneness.

Evidently, if anywhere, it is especially in definitions that the relation between subject and predicate may be called one of co-ordination: "Man is a rational animal," and convertibly, "A rational animal is a man." In other propositions the relation may be changed from superordinate to subordinate, according as we read them in either extension or in comprehension: "man is an animal," in the first case is interpreted, "man is a species under the class animal;" in the second,

---
[3] *The Principles of Logic*, cc. i. and ii.
[4] Zeller's *Socrates and the Socratic Schools*, c. xiii. p. 253.

"man, being a rational animal, includes animality under his total nature." "Extension," undoubtedly, is the aspect mainly chosen by Aristotelian logicians, who have good reasons for their preference; but we need not, therefore, deny that "extension" may fairly be said to have its basis in "comprehension." "Extension," better than "comprehension," could occasionally be dispensed with; for it is quite intelligible, though not necessary from all aspects, to teach with some logicians that an abstract term, such as "rotundity," has no "extension," inasmuch as it is a form prescinded from all subjects. On the other hand, an idea with no "comprehension" would scarcely be an idea at all—a point urged against Mill's doctrine that proper names have no "comprehension," or, as he says, "connotation." Carlyle's frequent use, in the plural, of words ending in -*ity*, may furnish examples showing, how to abstract terms an "extension" may be given: as when we predicate of several objects that they are each "lugubrities," or "fantasticalities."

After all, it is not the relative rank of subject and predicate, which is the vital point in the definition of judgment, but rather the copula. It is from the copula as centre that Aristotle, and St. Thomas after him, frame their definitions. According to the former,[5] "a simple proposition is the declaration that something *is* or is *not;*" it is "a synthesis of ideas, in which a truth or a falsehood is contained:" and, according to the latter authority,[6] "judgment

[5] *Prior Analytics*, Bk. I. c. i. n. 2.
[6] *Quæst. Disp.* quæst. xiv.; *De Veritate*, art. i.

is an act of intellect, whereby the mind joins or separates two terms through affirmation or negation." So defined, judgment is manifestly the act in which truth receives its completion; for it is in settling what are a man's *affirmations* or *negations*, what he says *is* and what he says *is not*, that we decide his correctness or error. Unless we can reduce his utterances to definite propositions, we cannot pronounce him right or wrong. While, however, we are thus considering the copula as specially decisive of the nature of judgment—as being the determining *form* to which the two terms serve as *matter*—we may, under another aspect, find the relation of *matter* and *form* repeated in the position of subject towards predicate.[7] For at least in what are called normal propositions, the subject stands for the whole thing in general, as it is in itself, while the predicate is some special form attributed to it by the mind; and the truth of the judgment is the truth of the application of this form. The very name "subject" signifies a recipiency of some determining form, not physically, but logically.[8] Thus in "aconite is poisonous," "aconite" stands for a whole object which the speaker might simply point out with his finger, or with a demonstrative pronoun: "poisonous" is a special notion which he has about the object, and he contends, that this notion rightly represents a determinate character

[7] Tongiorgi, *Logica*, n. 374.
[8] St. Thos., *Summa*, Pars I. quæst. xvi. art. ii. "Quando intellectus judicat rem ita se habere sicut est forma quam de ea apprehendit, tunc primo cognoscit et dicit verum."

or *formality* in the object, which formality he is now distinctly contemplating, and wishes to affirm. In some types of proposition this mode of interpretation will be less suitable, while in all the subject will be, not simply the thing in itself out of thought, but the thing ideally present in the mind of him who judges: else he could not judge at all. Still the point of view here indicated explains the use of the word "subject," and is of some assistance towards the attempt which has just been made, to give a definition of judgment,—an attempt the result of which may be finally stated in the few words of Cardinal Zigliara: "The act whereby we affirm or deny that a thing is." [9]

(*b*) But no sooner do we congratulate ourselves on being tolerably free from a troublesome question, than a philologist tries to drag us back into our old difficulties. It was all very well, he says, for Aristotle, St. Thomas, and others, who knew no language but Greek, Latin, and kindred tongues, to put the force of the judgment in the copula; but a wider range of linguistics brings the modern student across languages without the copula, and even without a verb strictly so called. As Mr. Sully urges, although our natural beliefs are expressed in propositional form, yet "progressing philology may show, that among many people confidence is really susceptible of expression in other than our affirmative forms of language." Nay even a melodic phrase on an instrument is declared, by Mr. Gurney, to be to

---

[9] "Est actus quo aliquid esse affirmamus aut negamus." (*Logica*, Lib. II. c. i. art. i.)

him, in more than a metaphorical sense, an affirmation. Reply is easy: all speech *equivalently* has the copula, even though this be not explicitly recognized. We ourselves, as children, once spoke with no conscious distinction of verb and noun: even still we occasionally omit the verb, or make a simple sound or gesture stand for a whole sentence.[10] Nevertheless every sentence, when rightly analyzed, is found to involve the sign of affirmation or denial. "There is," writes Max Müller, "beneath the diversity of human speech, that one common human nature, which makes the whole world kin. However different the families of language may be, so far as their material is concerned, let us not forget that their intention is always the same; and that if there are forms of thought common to all mankind, there must be forms of grammar too, shared in common by all who speak." More directly to the point is what Mr. Findlater writes in a note to James Mill's *Analysis:*[11] "Logicians, in treating of propositions, have almost exclusive regard to Greek and Latin, and the literary languages of modern Europe, which are all of one type. It might, therefore, be presumed that the theory thus formed would not be found to fit in all its parts, when applied to language of an altogether different structure. *The mental process must, doubtless, be the same,* but the words that express the several parts may be used in new and unprecedented ways." So obvious is this answer to a difficulty that it is scarcely necessary to insist further: but lest any one should be over much

[10] See a letter by Reid, given in *Hamilton's Reid*, p. 71.
[11] C. iv. s. 4.

moved by a plausible objection, the further confirmation of two more witnesses shall briefly be cited. These are the words of Mr. Sayce: "With all their differences the minds of most men are cast in the same mould. Thought is one, though the forms under which it shows itself are infinitely various. The unity which underlies diversity is seen in the tendency of all languages to assume common forms." Finally, Mr. Jevons shall speak: "Investigation will probably show that the rules of grammar are mainly founded upon traditional usage, and have little logical significance. This is sufficiently proved by the wide grammatical differences that exist between languages *though the logical foundations must be the same.*"

(c) No longer for the purpose of answering difficulties, but in order to shed more light on an important subject, a view taken by St. Thomas (1a, q. xvi. a. 2) with regard to judgment shall now be introduced, as eminently worth our study. He says that though, in an ordinary judgment, what we primarily assert is the fact, "This man is white:" yet indirectly we look to our own knowledge of this truth, not by a new act (*in actu signato*) but implicitly in the very act itself whereby we originally judge (*in actu exercito*). Each judgment is, as it were, accompanied with an "I know," or "as I perceive;" and but for this simultaneous consciousness of the rightness of our judgments, they would not have much intellectual value. For if to the vainglorious man it can be said:

> Your knowledge is nought, unless another knows that you know,[12]

[12] Scire tuum nihil est, nisi te scire hoc sciat alter.

much more may it be said to every man,

> Your knowledge is nought unless you yourself know that you know.

Now it is precisely this being aware that we know which characterizes a clear judgment, and makes it so confident, dogmatic, imperious. It bears its own inner conviction with it, as an indispensable condition; nor is this fact to be set aside for any mere theory, which asserts arbitrarily that one and the same act is incapable of attaining to self and to not-self. St. Thomas does not fall into the error which Mill lays to the charge of many Aristotelians, namely, that of supposing judgments to be about ideas instead of things: but he does insist on the important fact, which Mill also has noticed, that judgments are, as it were, lit up with a recognition of their own truth. It is this recognition which Mill[13] has in view when he says, that "belief" is the characteristic mark of a judgment. "It is impossible," he writes, "to separate the idea of judgment from the idea of the truth of a judgment; every judgment consists in judging something to be true. The element of belief, instead of being an accident, which can be passed over in silence and admitted only by implication, constitutes the very difference between a judgment and any other intellectual act. The very being of a judgment is something which is capable of being believed or disbelieved, which can be true or false, to which it is possible to say yes and no."

[13] *Logic*, Bk. I. c. v. § 1; *Examination of Sir W. Hamilton*, c. xviii pp. 347, seq. (2nd Ed.).

The last words admirably bear out the main thesis of this chapter, namely, that truth is specially in the judgment; but the passage also implies that consciousness of the possession of a truth is part of that possession itself. This consciousness, rather than the "readiness to act," on which Messrs. Bain and Clifford lay stress, is the mark of the judgment.

It is gratifying to find how different schools of philosophy confirm the doctrine of St. Thomas; but on this point, not to be diffuse, four very short illustrations, two German and two English, shall be the limit of quotation. Ueberweg[14] gives as the very definition of judgment the "consciousness of the objective validity of a subjective union of concepts:" while Bergmann teaches, that in judgment there is always conjoined with the apprehension of the object as simply existing, or as having these and those attributes, *a critical reflexion on the truth of these attributes, a verdict on the correctness of the attribution.* Of the English pair, Mill, whom we have just cited, further says: "The perception of truth or falsehood I apprehend to be exactly the meaning of an act of belief [a judgment] as distinguished from simple conception:" and Mr. Sully,[15] "Judgment is accompanied by a belief that the objects have a relation, or a relation corresponding to the relation in thought."

St. Thomas further supports his view by a contrast between intellectual judgment and mere sensitive, animal perception. "Though the sense can

---

[14] *Logic*, Part IV. parag. 67, et seqq.
[15] See the chapter on Judgment in *Outlines of Psychology*.

take cognizance of its sensation, it knows not its own nature, and, consequently, is ignorant also of the nature of its act and of its proportion to the object affecting it."[16] The lower animal can never take account of its own perceptions, whereas man recognizes himself as intelligent; the lower animal never recognizes truth as such, man does. Here again is a point which has so forced itself on rational observation, that representatives of the most widely divergent schools have a unanimity which, from their professed principles, might hardly be expected. In proof of the fact the only available method is quotation, but quotation shall be short, leaving each reader to make fuller verification for himself. After his own way of using words Lewes says, "To *perceive* a difference is one thing, to *know* a difference is another. The dog distinguishes meat from bread without *knowing* that one is not the other." Less explicitly Mr. Sully remarks, "An intelligent dog can distinguish and recognise, but he cannot mentally juxtapose objects, or compare them, except perhaps in a very imperfect and rudimentary way." It was from a like persuasion that a German philosopher declared his readiness to give a pig the honour due to a rational creature as soon as it intelligently affirmed, "I am a pig:" and another philosopher, of the same country, promised to dismount from his horse as soon as it said, "I am a

---

[16] "Quamvis sensus cognoscit se sentire, non tamen cognoscit naturam suam, et per consequens nec naturam sui actus nec proportionem ejus ad rem, ita nec necessitatem ejus." (*Quæstiones de Veritate*, quæst. i. art. ix.)

horse." The bacon for breakfast and the morning ride to digest it, are not much endangered by promises of this kind: for only a truly intelligent being, like man, can judge with full consciousness of the truth.

### Addenda.

Logicians, as it has been pointed out, can make an intelligible distinction between Apprehension and Judgment; but they leave over to psychologists a rather subtle piece of investigation as to the nicer discrimination of these two acts. How this inquiry has been pursued may be illustrated as follows:

(1) Whereas other writers largely tend to reduce Apprehension to Judgment, Hume would reduce Judgment to a case of Apprehension or Conception.[1] He regards it as "a very remarkable error," though one "universally received by all logicians," that the acts of the mind should be divided into Conception, Judgment, and Reasoning. "For, first, 'tis far from being true, that in every judgment we form, we unite two different ideas; since in that proposition, *God is*, or indeed in any other which regards existence, the idea of existence is no distinct idea, which we unite with the object. Secondly, as we can form a proposition which contains only one idea, so we may exert our reason without employing more than two ideas." As an inference needing no middle term, "we infer a cause immediately from its effect." The so-called three acts are reducible to one; "they are nothing but particular ways of *conceiving* our objects." The only note-worthy thing is *belief*, "which has never yet been explained by any philosopher," and leaves room for the putting forth of

[1] *Treatise*, Part III. § vi. note.

an hypothesis, namely, that belief is "a lively idea related to, or associated with, a present impression." " 'Tis only a strong and steady conception of any idea, and as such it approaches in some measure to an immediate impression."

(2) Reid[2] teaches, that in mature life a judgment goes along with every concrete apprehension. As regards abstract conceptions, he says indeed that apprehension may be exercised without either judgment or reasoning: but as he likewise teaches that in the perception, at least of sensible objects, the apprehension is derived from the analysis of the judgment, and not the judgment from the synthesis of mere apprehensions, he gives the absolute priority to judgment. "Simple apprehension, though it be the simplest, is not the first operation of the understanding; and instead of saying that the more complex operations of the mind are formed by compounding simple apprehensions, we ought to say that simple apprehensions are got by analyzing more complex operations."

(3) If Hamilton[3] and Mansel[4] are taken next, the reason is, not chronological order, but the fact that Hamilton's view appears in his Notes to Reid, and Mansel was a disciple of Hamilton. Hamilton finds fault with Reid, even for that degree of admission which the latter makes, when he allows that in case of abstract ideas apprehension can stand alone, without a judgment. "The apprehension of a thing, or the notion, is only realized in the mental affirmation, that the concept ideally exists, and this apprehension is a judgment. In fact all consciousness supposes a judg-

---

[2] *Intellectual Powers*, Essay I. c. vii.; Essay IV. c. iii.; Essay VI. c. i.
[3] See his notes on the above-cited passages from Reid.
[4] *Prolegomena Logica*, c. ii.

ment, as all consciousness supposes a discrimination. There is no consciousness without a judgment affirming its ideal existence." Hereupon Mansel distinguishes between psychological and logical judgment: "The psychological is a judgment of the relation between the conscious subject and the immediate object of consciousness; the logical is the judgment of the relation which two objects of thought bear to each other." Man judges psychologically when, as the idea "cow" passes through his mind, he simply recognizes the object as ideally existent—"there is a cow;" he judges logically when, for the terms of his judgment, he has two distinct concepts, "a cow is a ruminant." "The former cannot be distinguished as true or false, inasmuch as the object is only thereby judged to be present at the moment when we are conscious of it as affecting us in a certain manner, and the consciousness is necessarily true. The psychological judgment is coeval with the first act of consciousness, and is implied in every mental process, whether of intuition or thought. It cannot, therefore, be called prior or posterior to any other mental operation in which it does not take its place." Between judgment and conception Mansel's most concise distinction is that the two differ "in their data. In conception attributes are given to be united by thought in a possible object of intuition: in a judgment concepts are given to be united by thought in a common object."

(4) To go back now in chronological order we find that Dr. Brown[5] does not care much for the old traditional distinctions between apprehension, judgment, and reasoning: but rather insists on one great mental process, "relative suggestion," for putting all

[5] *Philosophy of the Human Mind*, Lecture li. Cf. Lecture xlv.

concepts into order, whether by judgment or by reasoning. "The tendency of mind, which I have distinguished by the name of relative suggestion, is that by which, on perceiving or conceiving objects together, we are instantly impressed with certain feelings of their actual relation. These suggested feelings are feelings of a peculiar kind, and require therefore to be classed separately from the perceptions or conceptions which suggest them, but do not involve them. . . . With the susceptibility of relative suggestion, the faculty of judgment, as that term is commonly employed, may be considered as nearly synonymous." Another passage bearing on the same point is one in which he compares what he calls perception and apprehension. "Simple perceptions are so feeble, dim, confused, and short-lived, and their objects are so numerous, run so into one another, come and go in such rapid succession, that the subject is unable to distinguish them one from another. . . . Perception becomes apperception by becoming more marked and distinct." This corresponds to a clear judgment. His reason for not using the more ordinary term "judgment" was given in an earlier Lecture:[6] "The term 'judgment,' in its strict philosophical sense as the perception of relations, is more exactly synonymous with the phrase I have employed (Relative Suggestion), and might have been substituted with safety, if the vulgar use of the term in many vague significations had not given some degree of indistinctness even to the philosophic use of it. Intellectual states of mind I consider as all referable to two generic susceptibilities —those of Simple Suggestion and Relative Suggestion. Our perception or conception of one object excites, of itself, and without any known cause external to the

[6] Lecture xxxii.

mind, the conception of some other object, as when the sound of a friend's name suggests the conception of himself: in which case the conception of our friend, which follows the perception of his name, involves a feeling of any common property with the sound which excites it, and might have been produced by the chair on which he sat, of the book which he read to us, &c. This is Simple Suggestion. There is another suggestion of a very different sort, which in every case involves the consideration, not of one phenomenon of mind, but of two or more phenomena, and which constitutes the feeling of agreement, disagreement, or relation of some sort. All the intellectual successions of feeling in these cases which constitute the perception of relation, differ from the results of Simple Suggestion in necessarily involving the consideration of more objects that immediately preceded them."

(5) Rosmini's[7] doctrine rests on his view as to the impossibility of deriving the idea of Being from experience: but, given this idea innately, it is what enables us to grasp our first conceptions of reality, and to grasp them by way primarily of judgments. In this sense he approves of Kant's doctrine, "that all our intellectual operations may be reduced to judgments, and the intellect generally may be represented as the judging faculty."

(6) Lewes[8] takes up something very like Brown's "relative suggestion" when he makes "grouping" the fundamental process of intellect. Each idea, as it comes up, groups itself with its likes, and marks itself off from its unlikes. The copula of the judgment is precisely this grouping. Every term is a judgment

---

[7] *Origin of Ideas* (English Translation), Vol. I. sec. i. c. iii. art. vi.; sec. iv. c. iii. art. xviii. xix. et alibi passim.

[8] *Problems of Life and Mind*, Vol. II. problem iii. c. ii.

completed and over: every subject is a group of predicates. The judgment lasts only while the grouping is being done: that once done, the judgment ceases to be and becomes a term. Mill and Bacon agree with Lewes that a proposition which has ceased to convey fresh information has become merely verbal, or, as Lewes words it, "a mere tautology."

(7) Mr. Spencer holds that nothing short of a "judgment" is an intelligent act; and if we take, as his description of "apprehension," the account which he gives of the formation of an "idea," we have the following account of it:[8] "It is because of the tendency which vivid feelings have severally to cohere with the faint forms of all preceding feelings like themselves, that there arise what we call *ideas*. A vivid feeling does not by itself constitute a unit of that aggregate of ideas entitled knowledge. Nor does a single faint feeling constitute such a unit. But an idea, or unit of knowledge, results when a vivid feeling is assimilated to, or coheres with, one or more of the faint feelings left by such vivid feelings previously experienced. From moment to moment the feelings that constitute consciousness segregate, each becoming fused with a whole series of others like itself that have gone before it: and what we call knowing each feeling for such and such, is our name for this act of segregation. As with the feelings, so with the relations between feelings. Each relation, while distinguished from various concurrent relations, is assimilated to previously experienced relations like itself. Thus result *ideas of relations*. What we call knowing the object is the assimilation of the combined group of real feelings it excites with one or more preceding ideal groups, which objects of the same kind

[8] *Psychology*, Part II. c. ii. § 373.

excited."[9] So much for the formation of ideas: and that these ideas are not mere apprehensions, exclusive of judgments, we are expressly told: "No state of consciousness can become an element of what we call intelligence, without becoming one term of a proposition which is implied if not expressed. Not only when I say 'I am cold' must I use the universal verbal form for stating this relation, but it is impossible for me clearly to think that I am cold without going through some consciousness having this form."[10] Below this stage of full intelligence he places a continuous process of evolution, starting from mere unconscious nerve-shock, gradually reaching sensation, and then, in the same smoothly ascending course, attaining successively higher points. "In the lowest conceivable type of consciousness, that produced by the alteration of two states, there are involved the relations constituting the forms of all thought." "In all cases perception is the establishment of specific relations among states of consciousness, and is thus distinguished from the establishment of the states of consciousness themselves. . . . Now the contemplation of a *special state* of consciousness, and the contemplation of the *special relations* among states of consciousness, are quite different mental acts—acts which may be performed in immediate succession, but not together. To know a relation is not simply to know the terms between which it subsists. Though, when the relation is perceived, the terms are instantly perceived, and conversely, yet introspection will show that there is a distinct transition of thought from the terms to the relation, and from the relation to the terms. While my consciousness is occupied with either term of a

---

[9] Compare Part II. c. viii. § 211; Part VI. c. xviii. § 355, and c. xxvii.
[10] Part II. c. i. § 60.

relation, I am distinguishing it as such and such, assimilating it to its like in past experience, but while my consciousness is occupied with a relation, that which I discriminate and class is the effect produced in me by transition from one term to the other."[11] By his whole treatment Mr. Spencer shows his great desire to make it appear, how from the simplest to the most complicate act of mind, the process is the same—a process which Hobbes calls "addition and subtraction,"[12] and Lewes "grouping." The passage in which Mr. Spencer sets forth the difference between perceiving terms and perceiving the relations between terms is considered by Mr. Guthrie to be one of the most important doctrines in the author's system: a doctrine, however, which Brown had before clearly enounced.[13]

(8) Under the present paragraph the reader need look for nothing more than a rough grouping together of authors who agree in the opinion, which may be usefully recurred to on various occasions, that the earliest judgments of the child are judgments in a very defective sense of the word. Very different minds concur in this observation, and herein lies the point of interest. Dr. Porter says, "The infant begins to perceive when, and so far as, it begins to attend. The soul of the infant is at first in a condition of activity, in which sensation greatly predominates, with only the feeblest exercise of intelligent perception. The infant at first feels many sensations, but it can scarcely be said to know objects at all; it perceives with the lowest activity possible of a power undeveloped by exercise." Perhaps it is something of the same sort which Luys, in his work on the brain, wishes to indicate when he writes: "Substantives play a principal part in the

---
[11] Part VI. c. xviii. § 354.
[12] *Leviathan*, Part I. c. v.    [13] *Human Mind*, Lecture xlv.

evolutions of thought and speech. They are the primordial data around which the verbs and other parts of speech group themselves. They are the elements that underlie the combinations of human thought." Again, Morell, in the *Outlines of Mental Philosophy*, expresses the opinion, that "both sensation and perception are prior to language. They cannot possibly be expressed in words, and conveyed to another. They belong to the more primary form of our intellectual activity."

From these non-scholastic authors we may turn to St. Thomas,[14] who speaks of two divisions of the sensitive faculty, which he calls *sensus communis* and *vis cogitativa*, and which, since he regards them as sensitive, he must conceive to be incapable of seizing an idea as such, reflexly and in its universality. They judge of concrete single facts, and serve as guides in individual cases. Now to the activity of such powers would often correspond those cognitions which Mr. M'Cosh, in his *Intuitions of the Mind*, talks of as preceding true judgment—cognitions that are "of the vaguest and most valueless character, till abstraction and comparison are brought to bear upon them." "An infant," says Mr. Sully,[15] "as an intelligent brute, may form a few rudimentary judgments, *e.g.*, *I am going to be fed*, without language. There may be many implicit judgments, where there is no statement. This applies to acts of perception and recollection. The child's first exclamation on seeing a large object, *big*, may be said to imply the statement, *that is a big object*. Singular judgments are the first to

[14] *Summa*, Pars I. quæst. lxxviii. art. iv.; *De Anima*, Part II. lectio xiii.

[15] See the chapter on "Conception" in *Outlines of Psychology*, where the author gives in detail Professor Preyer's observations on child life.

be formed by a child, and constitute a very important step in the development of thought." Mr. Sully's view of judgment proper has already been given, and need not be repeated here; but for the sake of marking an important distinction between sense and intellect, it must be noted that what he says about imperfect singular judgment, would, at least in many instances, be referred by the scholastics to what they call the *vis cogitativa*, and so far would fall outside the question of strictly intellectual acts. But even among these there must, at the beginning, be many mere dawnings of light, thin, vague, fleeting ideas, which just visit the consciousness, show a few of their connexions with other ideas, and then disappear.

To return once more to Mr. M'Cosh. He distinguishes "our primary cognitions and beliefs" from "our primary judgments," and builds the latter upon the former. "Every cognition furnishes the materials of a *judgment*, and a judgment possible, I do not say actual, is involved in every cognition. As the relation is implied in the nature of the individual object, and the judgment proceeds on the knowledge of the nature of the object, so the two, cognition and judgment, may be all but simultaneous, and it may be scarcely necessary to distinguish them except for rigidly exact philosophical purposes."

(9) According to Wundt, the content of the judgment is first given as an undivided whole, a whole which is not a mere bundle of associated ideas, but an apperceptive combination or *Gesammtvorstellung*. Judgment is the analysis of this whole, a dividing of it into parts as the very name *urtheilen* declares. Things first enter "into the field of view," and then "into the point of view:" the first is perception, the second apperception. The opposite theory supposes concepts first to exist

separately, and then to be put together by means of judgment.

(10) From the above list of opinions one obvious suggestion comes, that we ought not to be precipitate in drawing a very hard and fast line between apprehension and judgment, as between quite different acts of mind. The scholastics are prepared to recognize in the two a certain identity of act. If apprehension were taken, precisely on that side on which the intellect has to form its idea, at the suggestion of the sensitive image, the description of this aspect of the process, by the scholastics, may not seem to be allied to the description of a judgment. But if we take apprehension as they speak of it, no longer *in fieri* but *in facto esse*, no longer in process of being made, but as made, then, though the distinction be only mental and not real, it enables us better to understand how Suarez, after St. Thomas, teaches that the apprehension is a sort of judgment (*aliquale judicium*).[16] What Mansel calls the "psychological judgment" answers fairly well to the opinion of Suarez. An idea in the consciousness cannot be there, without affirming its presence and its object: it cannot rest simply in itself, as if it were a dead picture. It is a kind of cognition, and therefore tends to a judgment. Furthermore, when the mind is well stored with ideas, it is impossible that these should be present without in many directions asserting their mutual affinities; and so they stand, not as isolated concepts, but as more or less clearly formed judgments. When, however, two concepts are called up which, either in themselves, or at any rate for us, have no special connexion, they may remain in the mind with

---

[16] "Quatenus apprehensio est aliqua rei cognitio, est etiam aliquale judicium, quo implicite judicatur res esse id quod de illa cognoscimus." (*Metaphys.*, disp. viii. secs. 3 et 4.)

no tendency to enter into relation as terms of our judgment. Thus " Oxford eight " and " winners of the boat race " are complex terms, that may remain quite un-united by copula in the mind of an old oarsman, till he receives a telegram supplying the anxiously awaited " are " or " are not." Cases like this form, perhaps, the single exception to Dugald Stewart's law, that each mental state, as it comes up, asserts for itself a certain degree of credence—a doctrine re-affirmed by De Morgan, who, as he tells us, " takes it for granted that every proposition, the terms of which can convey any meaning at once, when brought forward, puts the hearer into some degree of belief." In using these words, he can hardly have had in mind the extreme cases of what are called *a posteriori* and synthetic propositions, in which the connexion of subject and predicate is a most purely contingent fact, the mere terms having no tendency to disclose a mutual relation in the shape of subject and predicate.

(11) Locke,[17] while fully agreeing with us that truth and falsehood are not properly in ideas, but only in propositions, yet has a peculiar use of the term "judgment," which calls for notice. He says:[18] "The faculty which God has given man to supply the want of clear and certain knowledge, in cases where this cannot be had, is judgment, whereby the mind takes its ideas to agree or disagree, without perceiving a demonstrative evidence in the proof, but presuming it." This is using " I judge " in the looser sense, which is obviously not the sense intended in the above discussion, as Locke himself would admit, who means by " proposition " what we have been signifying by " judgment."

As we have mentioned Locke, we may take occasion

---

[17] *Human Understanding*, Bk. II. c. xxxii.    [18] Bk. IV. c. xiv.

from his name to add some of Cousin's criticisms upon him, which bear directly on the priority between ideas and judgments, and are much in the spirit of some recent publications. " It is not true that we start with simple ideas, from which we proceed to those which are complex. Rather we begin with very complex ideas and proceed to those which are simple ; and the process of the human mind in the acquisition of ideas is the inverse of that described by Locke. Our first ideas are, without exception, complex, for the plain reason that all our faculties, or at least most of them, begin to act simultaneously. This simultaneous activity supplies us at one and the same time with a certain number of connected ideas, forming a whole. In a word, we have, at starting, a multitude of ideas which come to us contained or implied in each other, and all our primitive ideas are complex. A further reason for this is that they are particular and concrete."[19] Again, in the twenty-second lesson, he teaches that judgments are the primitive elements of thought, not simple ideas. " Language, that faithful expression of mental development, begins with compound propositions. A primitive proposition is a whole, corresponding

---

[19] "Il n'est pas vrai que nous commencions par les idées simples et qu'ensuite nous allons aux idées complexes; au contraire nous commençons par des idées complexes, puis nous allons aux idées simples; et le procédé de l'esprit humain dans l'acquisition des idées est précisément inverse de celui que Locke lui assigne. Toutes nos premières idées sont des idées complexes, par une raison évidente, c'est que toutes nos facultés, ou du moins un grand nombre de nos facultés, entrent à la fois en exercice; leur action simultanée nous donne en même temps un certain nombre d'idées liées entre elles, et qui forment un tout : en un mot vous avez d'abord une foule d'idées que vous sont données l'une dans l'autre, et toutes vos idées primitives sont des idées complexes. Elles sont complexes encore par une autre raison ; c'est qu'elles sont particulières et concrètes." (*Histoire de la Philosophie*, Leçon. 20me.)

to the natural synthesis by which the mind enters on the course of its development. These primarily formed propositions are in no wise abstract propositions, as, for instance, 'There are no qualities without a subject,' but wholly particular, as 'I am,' 'This body exists.'"[20]

[20] "Images fidèles du développement de l'esprit, les langages débutent non par des mots, mais par des phrases et des propositions très-composées. Une proposition primitive est un tout qui correspond à la synthèse naturelle par laquelle l'esprit débute. Ces propositions primitives ne sont nullement des propositions abstractes, telles que celles-ci : Il n'y a pas de qualité sans un sujet, pas de corps sans espace, et autres semblables, mais elles sont toutes particulières, telles que : J'existe, ce corps existe."

# CHAPTER III.

## DEFINITION OF CERTITUDE AND OF THE STATES OF MIND FALLING SHORT OF CERTITUDE.

*Synopsis.*
1. Definition of Certitude.
2. The question at present is one rather of definition than of fact.
3. Definitions of the states of mind which fall short of certitude.
   (a) Ignorance. (b) Doubt. (c) Suspicion. (d) Opinion.
4. Probability, a very large subject, not here discussed at any length.
5. The use of the word "belief."

1. THE assured possession of truth by the intellect is called Certitude, which is, therefore, defined to be the state of the mind when it firmly assents to something, because of motives which exclude at least all solid, reasonable misgivings, though not necessarily all misgivings whatsoever. The definition applies not only to every truth which is reached mediately by inference, but also to immediate intuitive truths, of which the motive lies simply in the self-evident connexion of the given terms. Hence it is not always needful to look for a motive outside of the judgment itself.

2. Such is a short description of what those competent to speak on the matter commonly understand by certitude. It is not yet formally under discus-

sion whether we mortals can arrive at such a state; though that we can is implied in every pretence to rational discussion of any sort. Still as far as explicit declaration is concerned, just as in an earlier chapter it was enough to say hypothetically, that if we have knowledge, it will bear a resemblance to the thing known; so now it suffices to say, that if we have certitude, it must be as above defined. Positively, however, to allow that we may, perhaps, be devoid of all certitude in our knowledge and that we must wait for philosophy to settle the doubt, this would be to cut from under our feet all available ground for philosophizing. But we may omit the explicit assertion of a fact without allowing it to be dubious.

3. Certitude is far from being our only mental condition in regard to things; and it becomes of the highest importance, for a well-ordered mind, to distinguish its several attitudes in relation to objects of knowledge. Some confused intellects make no attempt to sort their own contents, to put like with like, and to mark off the unlike by contrast; neither have such minds any clear views as to what they know or what they do not know. It would help them vastly, as the beginning of a re-organization, to note the following stages in the ascent from ignorance to certainty.

(*a*) Ignorance strictly so called, is either purely negative, simple nescience, or else it is privative,— want of some piece of knowledge which the person, all things considered, ought to possess. A surgeon need not know what the "eccentric" of a steam engine is, but he ought to know what a "tourniquet" is. Ignorance is not as bad as error; *per accidens*, it may even be "bliss;" but in itself at least it is no good, for it is nothing.

(*b*) Next to sheer ignorance comes doubt, which, in its widest sense, would include all the states intermediate between ignorance and certitude. But for technical purposes, or at any rate for the occasion, it is convenient to narrow down the meaning of the word by what in itself is rather an arbitrary limitation, and need not be borne in mind beyond the pages wherein the limitation is explained.

Mill[1] gives one definition of doubt which really belongs rather to sheer ignorance, when he describes doubt, "not as a state of consciousness, but the negation of a state of consciousnes — nothing positive but simply the absence of belief." It is true the scholastics speak of a *dubium negativum*, but they make it more than mere ignorance; they apply the term to the state of the mind we are in, when a question is proposed, and the mind, simply for want of any valid reasons on either side, remains quite neutral. Thus if we are asked whether some large assembly forms an odd or an even number, we lean to neither side, for lack of the means of deciding, even with probability, one way rather than another.

Now if, just for convenience, a name may be given to the perfectly balanced state, it can be called *negative doubt*, and comes very near to sheer ignorance; but is not quite sheer ignorance

---

[1] *Examination of Sir W. Hamilton*, c. ix. p. 133. (2nd Ed.)

because at least the question has been intelligently entertained, and its utter insolubility intelligently decided. It may be defined as the equipoise of the mind due to the absence of any valid reasons on either side. The parallel definition of *positive doubt* is "the equipoise of the mind, due to the fact that the reasons on either side are equal and opposite." In one case the balance is due to the absence of producible reasons, in the other case to the presence of exactly countervailing reasons. Of course it would be absurd to insist on the constant use of the words under these definitions, all the more so as no exact scales are usually at hand wherein to weigh reasons. Still the definitions are useful for the moment, while degrees between ignorance and certitude are being measured. Etymologically, according to Max Müller, *dubium* expresses literally the position between two points, and comes from *duo*, as *Zweifel* points back to *zwei*. The distinctions just drawn fit in well with the etymology.

(c) The first step out of doubt, when doubt is understood in the way above explained, may be called *Suspicion*, which is described as so faint an inclination to yield in one direction, that not even a probable assent is yielded, but there is a leaning towards a side.

(d) When, however, an assent is given, but as to a mere probability, and therefore only under restriction, there is *Opinion*, δόξα, if not quite in the Platonic sense, then in the general sense of what, from the appearance, seems likeliest, or at all events likely. In opinion, so defined, there is evidently wide room for variation between the limits of slender and of very substantial probability. It is a matter of choice whether we say that the assent is given to the probability of the proposition, or to the proposition as probable. Cardinal Newman, because of his special use of the word "assent," prefers the former expression. Again, the admission must be made, that in ordinary speech it would be absurd to insist on the use of the words "doubt," "suspicion," and "opinion," in strict accordance with the account of them just given; and yet the account has its manifest utility. It puts before the mind successive stages on the way to certainty, and gives to each a name. Now plainly it belongs to logic not only to treat of certitude, but also to compare it with other states of mind, which form the constant surroundings of our group of assured convictions. Only the intelligences that are blessed with the absence of all uncertainty can afford to confine their attention to certainty alone.

4. Much might be said of probability, but this is hardly the place in which to say it. Under certain aspects its treatment is largely mathematical; and as, in many instances, the mathematicians guarantee their results only for an infinite series, it follows that for any practical series they do not guarantee them to be strictly accurate. They cannot lay down any definite limits, however large, with the certainty that this will secure a fair game of chance, ending in a balanced condition.[2] For the definite period, say one thousand years, spent in

[2] Cf. Note A at end of chapter.

tossing heads and tails, may expire, just when a run of luck has fallen to one side. Still insurance companies, which, if no catastrophe happens, have a kind of interminable existence, can manage by statistics, not only to make their gains compensate their losses, but also give fair dividends to shareholders. For the information they require about the theory of chances, they look not to logicians, but to statisticians and mathematicians.

5. This chapter ought not to conclude without a remark on the use of the word "belief." To believe signifies sometimes (*a*) to hold a thing as a probable opinion ; and sometimes (*b*) to hold it as certain, whether (*a*) generally, without specially distinguishing the nature of the grounds or (*β*) specially, on the ground of the testimony of witnesses, or (*γ*) again specially, in cases where the object is not immediately presented to the perceptive faculties, *e.g.*, belief in a fact as remembered.

What Hamilton[3] says of belief may be usefully quoted as a help to the understanding of subsequent discussions in which his opinions will be involved: " Knowledge and belief differ not only in degree, but in kind. Knowledge is a certainty founded upon insight : belief is a certainty founded upon feeling. The one is perspicuous and objective, the other obscure and subjective. Each, however, supposes the other, and an assurance is said to be a knowledge or a belief, according as one element or the other predominates." Elsewhere he says,[4] " Belief is the primary condition of reason, and not

[4] Note A, on Reid, p. 760.   [3] *Metaphysics*, Lecture iv. p. 62.

reason the ultimate ground of belief." When, further, Hamilton teaches, that we believe the Infinite, yet cannot *conceive* it, or *know* it as possible, he does not wish to retract his declaration that what we *believe*, we must always, to some extent, likewise *know :* but he falls certainly into an appearance of contradiction : and beyond apology his views are at times misty and misleading. Perhaps it was some participation in them which prompted the line at the opening of *In Memoriam :*[5]

*Believing* what we cannot *know.*

The distinction is widely received, but probably not with very determinate meaning; sometimes it has its very legitimate sense of accepting, on sufficient authority, truths which we could not establish on their own intrinsic evidence, and which we do not fully comprehend after revelation.

With regard to the doubt which is often implied in the word "belief" it is, on religious principles, important to note, how the loss of dogmatic authority, and the assertion of private opinion, had much to do with spreading the erroneous notion that man's religious beliefs were but a set of opinions. Needless to say, in the Catholic Church, belief means *absolute certainty on the supreme authority of God.*

---

[5] "When I deny that the Infinite can by us be *known*, I am far from denying that by us it ought to be *believed.*" (*Metaphysics*, Lecture ii. Appendix.) As to how we can believe the inconceivable, see Mansel, *The Philosophy of the Conditioned,* pp. 69, 70.

## Note A.

On the theory of probabilities, and as to the fact that the very improbable sometimes happens, the following extract is instructive: "An extraordinary incident in a game at whist occurred in the United Service Club, Calcutta, a few days ago. The players were Mr. Justice Norris, Dr. Harvey, Dr. Sanders, and Dr. Reeves. Two new packs were opened and were trayed and shuffled in the usual way. Dr. Sanders had one of the packs cut to him, and proceeded to deal. He turned up the knave of clubs, and on sorting his hand found that he had the other twelve trumps. The fact was duly recorded in writing. The odds against the combination are 158,750,000,000 to one."[1]

[1] *St. James's Gazette*, February 14, 1888.

# CHAPTER IV.

### KINDS AND DEGREES OF CERTITUDE.

*Synopsis.*
    Preliminary remarks about the *assent* and the *motive* of judgment.
1. Species of Certitude.  (*a*) Metaphysical.  (*b*) Physical.  (*c*) Moral.
2. Degrees of Certitude.  Proofs of their existence: (*a*) From the side of objective truths varying in kind.  (*b*) From the experienced facts of compulsory, of easy, and of laborious assents.  (*c*) From the side of the subjective force of intellect, varying in different men.

*Addenda.*

BEFORE satisfactory advance can be made towards the next points of discussion, a few further remarks on the nature of judgment are quite indispensable.

It is a controversy amongst the scholastics,[1] which, as Cardinal Zigliara thinks, may, perhaps, be reduced, in the end, to a difference of words,[2] whether the assent in a judgment is completed in the clear perception of the relation between subject and predicate, or whether it is not rather another act, a sort of intellectual nod, following upon the perceived connexion of terms. One side says that the act of judging is itself a compound act, a compounding of

---

[1] Suarez, *De Anima*, Lib. III. sec. vi.
[2] *Logica*, Lib. II. cap. 1, art. i. § iii.

predicate with subject; the other side says it is a simple act, a simple affirmation or negation, following upon the comparison of the terms. The former party are careful to insist that there really is affirmation and negation, and they would not be content with any mere linking together or fusion of ideas, or any comparison, short of what is required by the meaning of the copula "is," or "is not." But, this asserted, they hold that no element of the judgment can be shown to be lacking when, in comparing the terms, the mind perceives, with or without additional light from outside the terms, the connexion between the two. In the following pages no distinct super-added act of assent will be supposed, on the ground that no argument in support of it seems convincing. If a man likes to confirm any of his judgments with a "Yes, that's it," the added act of approval is a new judgment, the result of reflexion on the previous one.

Taking, then, the assent—at least the legitimate assent—to be the perceived connexion between subject and predicate, we are able to reject a fallacious procedure which we must briefly describe. Those authors who make the assent a distinct act, following on the perceived connexion of the terms, occasionally manage to play some strange tricks in their account of a judgment, so that they can pronounce those propositions to be possibly doubtful which are generally reckoned indubitable. Thus Descartes, in behalf of his claim to be able to doubt certain mathematical truths, which seem indubitable, asserts, in explanation, his power to look away from the meaning or from the grounds of the proposition. He admits that while he considers the meaning of the terms he cannot doubt; but he contends that he can doubt as soon as he ceases to consider this meaning; and it is on this most flimsy pretext that he declares these truths to form possible objects of doubt :

"I have sufficiently explained, on several occasions, how this is to be understood. As long as the mind is attending to some truth of which we have clear conception, it is impossible for us to call it into question."[3]

It must be confessed that this passage, by suggesting the case of wilful, precipitate, and irrationally formed judgments, suggests also a most obvious argument in favour of judgment being a sort of simple nod of the mind, and not being intrinsically constituted by a perceived connexion between terms. The difficulty thus raised must stand over till we come to treat of the nature of error: and meantime it must suffice to say, that a solution is coming, and, further, that all error is *sub specie veri*; that it is because of some really perceived truth that the mind is able to assent at all; and that if the mind is carried on to add untruth to truth, or falsely to detract from truth, these are not strictly intellectual acts, but effects of obscurity in ideas, of the will, and of the force of association and habit. All which

[3] "J'ai assez expliqué, en divers endroits, en quel sens cela se doit entendre. C'est à savoir que tandis que nous sommes attentifs à quelque vérité que nous concevons clairement, nous ne pouvons alors au même façon douter." (*Méditations*, p. 467—Jules Simon's edition).

declarations must be expanded afterwards: at present it is enough to plead, that an assent worth the name cannot be wholly a sort of blind nod. In words anything may be said; but an assent not inwardly lit up with some intellectual motives is not strictly an intellectual act, for it is devoid of all insight. In a mixed act, the assent ceases to be intellectual at the point where insight ceases.

Let the word "motive" be clearly understood. The passage from ignorance to knowledge is a movement: therefore a motive power is required, one of the same order as the mind itself, an intelligible motive for an intelligent act. So far as any assent is not thus motived it is not properly an act of intellect. It is quite true that the intellectual faculty itself is a power to move, and the term motive might be used on the subjective side; but it is here regarded on its objective side, as an object soliciting the faculty, not as a faculty answering to the solicitation. In the proposition "A straight line is the shortest way between two points," the motive for the assent is intrinsic to the terms assented to—it is their own immediate evidence; in the proposition "A pistol-shot killed two recent presidents of the United States," the motive of belief is, with most of us, historic evidence, while with no one is it the intrinsic force of the two terms. In the one case, subject and predicate both terminate and motive the assent: in the other they terminate it, but do not motive. Many assents, the original motives of which are all, or most of them, forgotten, find still an adequate motive in the clear recollection that they have been validly established: at any rate, motive of some sort they must have present, under pain of being irrational.

1. The way is now clear for treating of the different *kinds* or *species* of certitude. In the terminology of Aristotelian logic, the species is what constitutes the essence of a thing : but certitude, in one respect, is not an essence, but only an accident of the mind. Essence, however, is an accommodating word, and allows of being varied in meaning according to the variability of ends in view; the same difference becomes essential or non-essential, specific or non-specific, with the change of purpose. A round biscuit, and a square biscuit, both of the same material, differ specifically for the mathematician, non-specifically for the child who eats them. In general, according to the usage of human speech, that difference in any order is to be regarded as specific, which, in relation to that order, goes to the very essence or nature of the thing. If we want a red object to excite a bull, then the colour, not the material, is the specific character: if we want a woollen garment, then the material, not the colour, is specific.

This being so, we observe that the essential character of certitude, that which radically distinguishes it from other states of mind, such as suspicion or opinion, that which gives it its lower generic place under the higher genus of intellectual assents, is the firmness of the assent. In other cases we either withhold assent, or give it only with reserve; but in certitude we are without any doubt.

In the firmness of the assent, therefore, if anywhere, specific differences of certitude are to be found; for differences here will be difference within the essential constituents. In establishing three such differences we shall be disregarding one pet modern theory about the non-necessary character of all truth; but it will be better to go on our way in disregard of adversaries for the present, and to come back again, in the next chapter, to see what objectors have got to urge.

(*a*) The highest motive of certitude, giving the highest species of assent, is *metaphysical;* which implies a necessity so absolute as to be bound up with the immutable nature of God Himself. In this sense we may adopt, or rather adapt, the heathen saying, ἀνάγκῃ δ' οὐδὲ θεοὶ μάχονται; God cannot fight against the necessities of His own all-perfect Nature, and their inevitable consequences in regard to the possibilities of creation. But we must avoid the pagan error of looking upon this necessity as something extrinsic to God, a fate or destiny having an independent existence. The prime metaphysical necessities are that God should exist as the one absolutely necessary Being; that He should be just what He is, and that from the nature of this First Being should follow the laws regulating the possibilities of all that can be created, and of all finite truths. This is the matter of a whole section in the scholastic treatise on General Metaphysics. Here it must suffice to give a few specimens of truths metaphysically necessary; to which the ordinary mind, unbiassed by philosophic theory, will feel no difficulty in allowing a most absolute, irreversible character. "God cannot lie;" "Moral right is sacred;" "Nothing can at once, under the same aspect, be and not be;" "Every new reality or event must have a cause;" "Two straight lines cannot enclose a space." These examples at least give a clue to what is meant by the species of truth called metaphysical.

(*b*) Strongly contrasted with absolute necessity is *physical* necessity, which we call contingency. The physical ($\phi\acute{v}\epsilon\iota\nu$) is what comes into being, what has an origin and a growth, what is produced or made; and so it differs from the metaphysical, which simply and eternally is. Hence the term physical, as here understood, does not apply to the order of mere possibilities.[4] The physical is the actually created order of real existence, which existence is contingent, and might never have been. As a fact, out of the various worlds which in His Omnipotence God might have created, He has created the existent universe; whereas He might have created another, or might have abstained from creating altogether. Even the present system is not so rigorously settled that He cannot miraculously interfere with the ordinary sequence of effects. Thus the physical necessity, to which we have to bow, is not *a priori* and immutable, but *a posteriori*

---

[4] This use would exclude God from the order of the "physical," though as far as He has real existence He is often included under it; the fact being, that "physical" is a term of varied meaning, as when we distinguish physical science from the science of things spiritual, physics from chemistry, physics from physic, and so forth.

and mutable by Divine power. It is, however, *a priori* to this extent, that all its possibilities were fixed *a priori*, and to an intellect able to look into the very constitution of bodies, all their powers would presumably be thence deducible; while from the primitive collocation of world-elements, all subsequent phenomena, apart from what is due to the interference of free will, might be calculated. We, however, who can neither adequately penetrate the inmost nature of matter, nor quite solve even the comparatively simple problem of three attracting bodies, have to proceed on a humbler method: so for us physical truths are *a posteriori*, and are ultimate facts, which we take on the evidence of experience, without being able to give their final account. All our physical explanations end in mere empirical facts.

(c) The third kind of motive for certitude need not detain us long; for we shall have to give a separate consideration to historic evidence, and then the nature of *moral* truth, as it is styled, will appear in a fuller light. We have, in this matter, the difficult problem to find a sort of necessity, in spite of free will being mixed up with the elements of our calculation. We shall have to claim, that occasionally we can know how people have used, or will use, their power to choose under given circumstances. Thus moral truth, in the sense at present given to it, is truth about human action, which in many details is free, though it has not a freedom unbounded. The theoretic difficulties against the possibility of ever calculating human conduct on the hypothesis of a real liberty of choice, vanish at once before a concrete case; it is a sheer impossibility that historians should be deceiving us, when they narrate certain substantial events in the lives of Alexander the Great, Julius Cæsar, and Charlemagne. Again, there are cases where we may be certain, that a very well tried character will not prove treacherous under moderate temptation and enormous responsibility. These instances convey sufficiently what is to be understood by the third species of certitude, which is here styled moral.

It must be added, however, that the same phrase, "moral certitude," which is here used for strict certitude, is employed also in a looser way to mean high probability, such as would be enough to determine the action of an ordinarily prudent man. It is moral inasmuch as it suffices for a moral agent. Thus a merchant would make a great venture on "a moral certitude," which meant the probability of a thousand to one, yet did not quite leave the level of probability, and mount into that of strict certitude.

While it belongs to philosophy to draw, in general, the distinction between the three species of certitude, it would be preposterous to ask it to settle, in all concrete cases, whether we can have certitude, and if so, of what kind. Not all the departments of science together can discharge this function: but each department is left by philosophy to do what it can in its own sphere, while philosophy itself investigates certitude in its highest generality. Should any one try to illustrate its doctrines by examples confessedly dubious in their own scientific order, it simply begs the person to choose a more suitable illustration ; it does not undertake to meddle out of its own province, and it borrows, to exemplify its teaching, only safe instances.

2. Next to difference of kind in certitude may be taken difference of degree. It is maintained by some, that from its very nature certitude admits of no degrees, of no less and more; that if a man is sure, he is sure, and that is all about it ; so that to talk of assurance being made doubly sure is a mere *façon de parler*. We are not concerned to maintain that under no acceptation of the word "assent" is it possible to deny the existence of degrees in it; but if we take "assent" in its wider and more ordinary meaning, the certitude of our assent does admit of degrees, in the sense we are about to explain in the following paragraph.

Every certitude must absolutely exclude all solid doubt, which exclusion of doubt is the negative side of certitude and, of its own nature, allows of no degrees ; but the positive side, or the positive assent itself, is of a nature to admit degrees. Certitude, then, on its *negative* side has not, on its *positive* it has, degrees. The two sides are only distinguishable aspects, not separable elements ; by one act, we are sure and do not doubt. Before this doctrine can appear quite satisfactory, it needs a little elaboration. For against it, in its cruder form, might be urged the fact, that the same motives which produce assent also drive out doubt ; and that, therefore, doubt is expelled with a force varying with the expelling motives, all of which are alike in that they annihilate doubt, but differ in that some effect the annihilation with greater energy. So of two men who agree in the fact of being no longer in a certain assembly room, one may have been quietly lifted into the adjoining street, and another shot with a catapult into a street some distance away. Thus, it is argued, even the negative side of certitude, the expulsion of all doubt, may differ in degree.

To answer the objection we must limit the meaning of expelled doubt. We must take the absence of doubt purely on its negative side, or on its side of nonentity; then nonentity, as such, is unsusceptible of less and more. Of course mathematicians, with whom, however, positive and negative often mean no more than one direction and its opposite, extend negative quantities as far backward as positive quantities go forward: to the *plus* series 1, 2, 3, 4, &c., they can oppose a *minus* series 1, 2, 3, 4. Still it will be only by positive considerations that degrees are estimated in the negative direction. For example, a man who is said mathematically to be *minus* £1,000, is interpreted to have no money, and, worse than that, to be under obligation to give the first available £1,000 he gets to his creditors. Here the negative, as a negative, is the fact of a man having no money: beyond that the degree of his indebtedness must be calculated on positive grounds. The case is not quite parallel with the one in hand; but it sheds upon it some light, by helping to show how a negative, as a nonentity, cannot be greater or less. A negation in the sense of an intellectual denial may be given with greater or less intensity; but a negation in the sense of the mere non-existence of a doubt has no varying intensity. And so the whole statement, that certainty, on its negative side, has no degrees, is reduced to saying, that the non-existence of doubt in every certitude is a simple non-existence, or nothing; and that nothing does not admit of more and less.

The way is thus cleared for establishing the possibility of degrees on the positive side of certitude. What a man *does* is one thing, what in strict logic *he ought to do* is another; and speaking from the former point of view only, it is not incumbent on us to prove that every man *does* always regulate the degree of his assent according to the considerations now to be brought forward. Cardinal Newman ably maintains that as a fact man does not so proportion his assents; it is enough for us that the considerations we have been urging are such that, of their own nature and *cæteris paribus* they produce the effect of varying the force of intellectual adherence. Those who, like Dr. Gutberlet, start from the notion that certitudes are equations,[5] and argue that however the terms equated may vary, yet equation itself is constant, plainly leave out of the question elements which claim to be noticed.

St. Thomas, while he allows to the objection derivable from this idea of knowledge as an equation, the truth it contains, still manages to take out of the difficulty all its force as an objection.

[5] *Logik*, Die Erkenntnisstheorie, I. 4.

Not, indeed, that he is designedly combating the adverse view: his words form only the better answer to the difficulty, because they meet it unintentionally and in the mere act of explaining the real position of assent. What he says is this: "According as a thing happens to have more truth in it, it elicits a higher belief. For while truth consists of an equation between intellect and object, if we regard truth merely as an equation, it does not allow of less or more: but if we consider the very Being of the object, and remember that truth has its ground in the Being, and that such as the Being is, such is the truth; then those things which have more of Being have also more of truth."[6] That is, if you regard truth as a mere equation between a mental act and its formal object, equality is equality all the world over, whether the terms equated be greater or less. But intellectual assent is no mere dead uniform sign of equality, like our algebraic symbol: it is a living response to objective evidence, and is apt to vary, *cæteris paribus*, with the evidence that calls it forth. Hence St. Thomas again affirms: "An assent is nothing but the determination of the intellect to one affirmative: and by so much greater is the certitude by how much stronger is the motive by which it is determined."

Arguing first of all on the line here suggested, we may hope soon to find force of demonstration enough to overpower the hostile statement of Mr. Lewes: "The widest of all axioms, *whatever is, is*, cannot be more certain, more irresistible, than the

[6] St. Thos., *Quæst. Disp. De Caritat.*, art. ix. ad 1.

most fleeting particular truth." Against this let us try three arguments.

(*a*) Whilst certitude always remains up to the level of certitude, and never sinks to the lower grade of strong probability, still its accidental degree may vary: it reasonably so varies when the truth proposed is of a higher order. Thus a man is certain that he lit his fire with one of Bryant and May's matches: he is, or may be, certain in a more intense degree that the fire would not have blazed up without an igniting cause of some sort. He is certain that Victoria in 1887, the Jubilee year, was Queen of England: he is, or may be, certain in a more intense degree that God is sovereign Lord of all. He is certain that he paid a bill for four shillings with two florins: he is, or may be, certain in a more intense degree that two and two make four. Where an adequate cause for intenser degree is assigned, and where we have a faculty susceptible of stronger and weaker excitation, it is fair to infer a possible variation in the effects. To say that the variation is something outside the rational assent, or that it belongs only to concomitant emotion, is to ignore explained facts. It is quite true that some degrees of intensity are emotional; as when an Englishman assents more keenly to the authentic news of a victory for the British arms, than to the equally authentic news of a victory gained by one savage tribe over another. But the possibility of degrees in the region of emotion does not exclude their possibility in another. It was in intellectual motives that a cause for intenser assent was above pointed out; and therefore it is in intellectual assent that the intenser degree may sometimes reside: which is all we had to show.

(*b*) A second argument may be put thus : Always supposing true certitude, sometimes we assent as under compulsion, and perhaps against our wish to believe otherwise : sometimes we assent, with ease indeed, but not with the feeling of strong compulsion ; sometimes we assent but not without a certain effort of the will, urging on the mind to put carefully together and admit the just sufficient evidences. Against saying universally that these represent three descending grades of assent stands the fact, that the firmest of all assents, the act of supernatural faith, results from a command of the will; but keeping within the natural order, and speaking of general cases, we may assert of the above, that they are three varying grades, the variation being precisely in the intellectual character of the acts.

(*c*) Again, if the simple argument, "Certitude is certitude all the world over," were decisive of the whole question, it might be questioned whether Divine and Angelic intelligence were to be regarded as following under the rule. But waiving these points and keeping strictly to human intelligence, just as we drew a proof from the varying force of objective truths, so we may draw a proof from the varying force of human minds ; some men, because of their keener faculties, may give an intenser assent to the same argument which draws likewise the assent of their duller brethren. And so once more, a certitude can vary in degree.

## Addenda.

(1) In distinguishing three kinds of certitude it is worth while to notice their interdependence. Though some physical conditions of brain must be fulfilled in order that the mind may understand a metaphysical truth, yet man may claim, in regard to metaphysical truths, that he can obtain them without the admixture of truths of a different order. The principle of contradiction is reached in its purely metaphysical character. But all physical truth must be inseparably bound up with some metaphysical principles; for example with the just-mentioned principle of contradiction. For where would be the use of discovering that a planet exists, if there were no guarantee that its existence was incompatible with its non-existence. Obviously the metaphysical principle is not applied after the physically ascertained fact, but enters indissolubly into union with such ascertainment, which else would be impossible. In the third place moral truth must have joined with it, not only metaphysical, but also physical truths: for we judge human conduct through physical manifestations, and human speech or writing is equally a physical phenomenon.

(2) It has been asserted that the intenser degree is never in the certitude as such, but in some concomitant emotion. Thus a writer in *Mind*, who betrays the fact that he has not cleared up his own thoughts on the subject, ventures on the declaration, that "there are no degrees of intensity in cognition: the intensity is a matter of feeling concomitant with the cognition."

The relation between what are called feeling and

[1] See Mr. Bain's *The Senses of the Intellect*, Introduction, c. i.; Mr. Spencer's *Psychology*, Part IV. c. viii.; Lotze's *Microcosmus*, Bk. II. c. ii.

cognition forms a matter of much vague discussion.[1] Some place the foundation of feeling in cognition, on a wide extension of the principle, " There is neither desire nor fear of the unknown ;" others reverse the position, and make blind feeling primitive—pure subjective feeling without an object. Feeling again is made to include all consciousness ; so that a stronger intellectual assent, making itself felt in consciousness, could be put down to feeling.

In face of such ill-defined terms in the objection, it is enough to reply, that if, from an examination of any case, it appears that one assent has no *intellectual* motive or cause stronger than another, then it is no illustration of our thesis ; but if a distinctly intellectual ground of superiority can be shown, then it is an illustration. If exactly the same vouchers tell a man of the equally credible events that a friend and a stranger have both perished in a shipwreck, then the intenser act in regard to the friend's death may be put down to emotion. There may also be something intenser in the intellectual energy, but this element would be difficult to detect and estimate.

And generally we may say, that people are too apt to think that they can mark off, with nicety, assent from assent, affective movement from affective movement, and the former of the two elements from the latter. Whereas clearly to isolate an act in reflexion, is often most difficult or impossible. It may very well be that, not acting on the possible principles explained in the argument which we used to prove the greater force of a metaphysical over a physical truth, a schoolboy will concentrate even a greater intellectual energy on the very contingent fact that he, a poor player usually, has had the luck once to score fifty at a cricket match, than on the eternally abiding, necessary truth that two straight lines cannot enclose a space. But in a concrete case of this kind, who is to disengage the intellectual from the emotional elements? Again, the mere size or amplitude of the object assented to may easily get confused with a notion that the assent itself is intenser; and who then is neatly to discriminate extension from intensity? Take once more the rule sometimes laid down, that in any given case feeling and intelligence are in inverse ratio; the heavier drain in one direction exhausting the supply in the other. There is some truth expressed in such a rule; but on the other hand, the force of an emotion is sometimes to increase the intellectual power, not to diminish it, as in those who speak best under a fairly strong excitement. Let us not, then, be deceived by a fancied simplicity, but rather apply to acts of the human soul what a French writer, quoted by Sir H. Maine, says of human society: " I have hitherto discovered but one principle which is so simple as to appear childish, and which I scarce dare to express; it is no other than the observation, that a human society, a modern society especially, is an immense and complicated object."[2]

A human intelligence too works by a very complicated process.

[2] "Jusqu' à présent je n'ai guère trouvé qu' un principe si simple qu'il semblera puéril, et que j'ose à peine l'annoncer. Il consiste tout entier dans cette remarque, qu'une société humaine, surtout une société moderne, est une chose vaste et compliquée."

# CHAPTER V.

## METAPHYSICAL AND PHYSICAL CERTITUDE
## (*Continued*).[1]

*Synopsis.*
  I. Metaphysical certitude.
     1. Mr. Huxley's three meanings of necessary truth. (*a*) Uniformity or consistency in the use of terms. (*b*) Indissoluble association. (*c*) Facts of immediate consciousness.
     2. Argument in behalf of metaphysical truth from the admission of adversaries. (*a*) Admissions as to moral truth. (*b*) Admissions as to intellectual principles.
     3. Defence of metaphysical certitude.
  II. Physical certitude.
     1. The sum total of matter and force a constant quantity. Various meanings of "the Uniformity of Nature." (*a*) Like agencies, under like conditions, will always have like effects. (*b*) The sum total of physical agencies in the world is constant. (*c*) Nature presents periodic phenomena, or the recurrence of like events in her course.
     2. Physical science saved on principles above enunciated, lost on principles of pure empiricism.
     3. Distinction drawn between simpler physical truths, on which we can have certitude, and more complex, on which often we cannot have certitude.
     4. How to judge that no miraculous interference need be suspected.

SOME years ago, what has been briefly laid down about metaphysical and physical certitude would have been much more readily taken for granted than

[1] Beginners may omit this chapter.

it will be to-day, when so many are boasting that they have changed the prevalent ideas on the subject. It will be the endeavour of this chapter to show that the change is not for the better, and to recommend a return to the old way of thinking.

I. Starting from the examination of metaphysical truth, we must carefully guard against a prejudice, with which some seek to discredit the cause; the notion, namely, that those who hold some principles to be in a real sense *a priori* and beyond mere experience of facts, are thereby committed to the assertion of innate ideas.[2] This is not so. They allow that all human knowledge is started by experience, internal or external; but they further contend—and here they differ from *pure empiricists*—that while some truths might have been different, other truths are perceived to be founded on absolute necessity, and are therefore valid for all places and for all times, nay, even beyond all place and time. In the latter case, though our knowledge has its origin in single experiences, yet no sooner have the ideas been grasped, than they are seen to imply universal principles.

1. To understand against what manner of teaching we have to contend, it will be well to examine the three meanings, which Mr. Huxley,[3] in his little work on Hume, thinks it possible to attach to "necessary truth."

(*a*) The first interpretation is founded " on the convention which underlies the possibility of in-

[2] See Mr. Bain's *Mental Science*, Bk. II. c. vi. n. 1.
[3] C. vi. See also Mr. Bain, loc. cit. n. 7.

telligible speech, that terms shall always have the same meaning." This is what Mr. Bain, an expounder of the philosophy which Mr. Huxley substantially adopts, has called "the principle of consistency,"[4] which he thus formulates: "It is a fundamental requisite of reasoning, as well as of communication by speech, that what is affirmed in one form of words shall be affirmed in another." The need of this rule no one will deny, if he wishes to secure intelligible communication between men, whose principal means of intercourse is by speech. But, while needful, the rule holds a very secondary place in the philosophy of the subject; for, deeper than consistency of speech is consistency of thought, and deeper than any mere consistency of thought is its correspondence to the reality of things. Now this correspondence, neither Mr. Huxley nor Mr. Bain attempts to defend; they reject the definition of truth, as "conformity of mind to thing," inasmuch as they both proclaim that idealism cannot indeed be proved, but neither can it be disproved.

On the matter of this all-important consistency of thought with things Mr. Bain[5] has to content himself with making three postulates, one for objects of present consciousness, another for objects of memory, and a third for objects of expectation in the future. On the first point "we must assume that we feel what we do feel; that our sensations and feelings occur as they are felt. Whether or not

[4] *Mental Science*, loc. cit.
[5] *Mental Science*, loc. cit.; *Inductive Logic*, Bk. II. c. ii.; *Deductive Logic*, Appendix D.

we call this an irresistible belief, an assertion whose opposite is inconceivable, we assume it and proceed upon it in all that we do. Calling its negative unthinkable does not constitute any reason for assuming it: we can give no reason better than that we do assume it." Secondly, belief in memory is also, and more especially, taken as a practically needful assumption for which we can assign no reason in justification. And thirdly, to crown the whole work of assumption, and to do away with all solid motive for trust that our thoughts represent things, the two first postulates are supplemented by a third, and not only supplemented by it, but made in some sort to rest on it for support; at least there is a reciprocal dependence between the three. "What has uniformly been in the past," says the third postulate, "will be in the future; what has never been contradicted in any known instance, there being ample means and opportunities of search, will always be true." For this postulate, "we can give no reason or evidence:" indeed it is "an error to give any reason or justification," instead of treating it as "begged from the outset." At all events, "if there be a reason it is practical and not theoretical;" theoretically or rationally considered, the postulate "involves a hazard peculiar to itself, and any belief as to the future which we adopt on its authority is " a perilous leap." Nay, experience is even positively against the postulate, testifying to us that "nature is not uniform in everything," by the "establishment of exceptions to uniformity." So situated, "we go forth in a blind faith until we receive a check. Our confidence grows with experience, yet experience has only a negative force; it shows us what has never been contradicted, and on that we run the risk of going forward on the same course." Furthermore the curious fact is noted, that, although without justification for itself, "this assumption is ample justification of the inductive operation, as a process of real inference. Without it we can do nothing, with it we can do anything."

The passages thus quoted have an immediate bearing on physical truth, in relation to which we shall presently consider them; but they have also a connexion with metaphysical truth, on which account they have been thus early introduced. The connexion is this: we are speaking of metaphysical truth, another name for which is necessary truth. Now the first meaning assigned by Mr. Huxley to necessary truth is "consistency of language." Even if we suppose this consistency of language to be backed by a corresponding consistency of thought, we may not suppose, without inquiry, that behind the consistency of thought there is secured a solid basis of objective reality. Investigation shows us that such foundation is not secured; as well because of Mr. Huxley's own assertion that idealism cannot be disproved, as because of Mr. Bain's futile attempt to rest the objective reality of thought, for past, present, and future, on three postulates, of which he gives a most lame account. They are three postulates in the worst sense of question-begging. We conclude, therefore, that the first of the three suggested meanings of necessary truth is quite inadequate. To repeat once more and emphasize the main burden of complaint, the school to which Mr. Huxley has attached himself, does not make any provision for a knowledge of necessary truth *about things*. Just as Mill declares that he cannot extend the principle of contradiction *to things* in themselves, nor absolutely make of it more than empirical law of our thought, so Mr. Bain similarly stops short of reality. " Were it admissible," he writes, "that a thing could be and not be, *our faculties* would be stultified. That we should *abide by a declaration* once made is indispensable to all understanding between man and man. The law of necessity in this sense is not the *law of things*, but an unavoidable accompaniment of the *use of speech*." So explained, the law is quite empty of reality.

Yet inadequate as it is, Mr. Bain does not allow it its full force. He mentions as being outside the range of consistency in speech or of "truths of implication," the axioms that "things equal to the same thing are equal to one another," and that "the sums of equals are equals;" also the principle that "every event must have a cause." These several propositions, he maintains, are reached inductively, are "not necessary," and "may be denied without self-contradiction." So much for necessary truth when described in Mr. Huxley's words, as "the convention underlying the possibility of intelligible speech, that terms shall always have the same meaning."

(*b*) Let us try the second interpretation of

necessary truths; now they are "propositions the negation of which implies the dissolution of some association, memory, or expectation, which is in fact indissoluble." Fastening on the word "association" we have one of the terms round which so much of the present controversy gathers; nor is it possible intelligently to conduct the discussion unless we understand the large part played in the philosophy of our English empiricists by association. In this matter Mr. Huxley often follows so closely the footsteps of Mill, that it is better at once to recur to the more original author, though Hume most deserves to be called the prime offender.[6]

Mill, however, is not such an out-and-out associationist as it might, from some of his utterances, appear. It is true that not only in intellectual processes, but even in volitional, he attributes very much to association. Denying free will, and yet clinging to what might easily be taken as a remnant of the belief in freedom, after a manner which it puzzles even his friend, Mr. Bain, to regard as other than an inconsistency,[7] he was alarmed, at one

---

[6] *Treatise on Human Nature*, Part I. § iv.

[7] His theory is, that though man's conduct is rigorously determined by character and circumstances, yet man can do something to improve his character. "Modified fatalism holds that our actions are determined by our will, our will by our desires, and our desires by the joint influence of the motives presented to us and of our individual character; but that our character having been made for us and not by us, we are not responsible for it, nor for the actions it leads to, and should in vain attempt to alter them. *The true doctrine* maintains that not only our conduct, but our character, is in part amenable to our will; that we can by employing proper means improve our character."—*Examination*, c. xxvi. p. 516. (2nd Ed.)

period of his life, lest his early educators should not have formed in him associations of right conduct sufficiently strong to keep him always on the line of rectitude. But it is on the intellectual side of association that we are at present considering his views; and here he distinctly departs from his father's teaching, that judgment is mere association.[8] He declares that belief is a new element of a special kind, though he nowhere goes so far, as does Mr. Bain, in the assertion of spontaneous beliefs, exceeding all warrant for their formation. According to Mr. Bain:[9] "it may be granted that contact with actual things is one of the sources of belief, but it is not the only nor the greatest source. Indeed so considerable are the other sources as to reduce this seemingly preponderating consideration to comparative insignificance." Mill rather adheres to the view, that in producing belief the force of association is at least preponderant, as will be manifest in instances now to be adduced.[10]

He divides indissoluble associations into those which we cannot so much as conceive to be reversible, and those which he fancies he can conceive to be reversible; but not even the former will he pronounce absolutely irreversible. For "it is *questionable*," he holds, "if there are any natural inconceivabilities, or if anything is inconceivable to us for any other reason, than because

---

[8] See his note to James Mill's *Analysis*, c. xi. n. 98.
[9] *Logic*, Introduction, n. 7, Bk. VI. c. iii. n. 1.
[10] *Examination*, c. vi. pp. 67, 68. (2nd Ed.)

nature does not afford us the combinations necessary to make it conceivable." More strongly still, passing from the phrase, "questionable," to "can only be," he says, "If we have any associations which are in practice indissoluble, it *can only be* because the conditions of our existence deny us the experience which would be capable of dissolving them."

After such declarations we are not surprised to find how ready Mill is to allow the possibility of dissolution in associations which, he says, are to us at present, not alterable in any form that we can conceive. Apparently forgetful of his admission that judgments are more than associations of ideas, he takes, as test cases, the three primary principles of identity, contradiction, and excluded middle; and about them he avers,[11] "I readily admit that these three principles are universally true of all phenomena. I also admit, that if there are any inherent necessities of thought, these are such. I express myself in this qualified manner, because whoever is aware how artificially modifiable, the creatures of circumstance, and alterable by circumstances, most of the supposed necessities of thought are, (though real necessities to a given person at a given time), will hesitate to affirm of any such necessities that they are an original part of our mental constitution. Whether the three so-called fundamental laws are laws of thought by the native structure of the mind, or merely because we perceive them to be universally true of observed phenomena, I will not positively decide; but they are laws of thought now and

[11] *Examination*, c. xxi. p. 417. (2nd Ed.)

invincibly so. They may or may not be capable of alteration by experience, but the conditions of our existence deny us the experience which would be required to alter them." This passage is the plain negation of all certitude; for if with regard to such self-evident truths as that "whatever is, is," and that "whatever is, cannot at the same time, and under the same respect, not be," we are unable to rely upon our clear mental insight when it tells us that these axioms are true for all intelligence and beyond all possibility of alteration; then we never can have any really solid foundation for a firm assent. Certitude even ceases to have a meaning.

To pass now to those metaphysical truths which Mr. Mill thinks to be conceivably alterable, under conditions of experience other than what this world affords; we will take his assertion, that to beings differently situated square-circle might be as rational as sweet-circle is to us. His argument is, that just as to us the sensations sweet and circular may be derived together from one object, so to persons of another constitution, or in other surroundings, the sensations of square and circular might be derived together from one object. It is a revelation of the thorough unsoundness of Mill's philosophy, when he thus confounds sensations with intellectual perception of universal truths. So long as he looks only to chance association of sensations, he may fancy that any combination of these is possible; but if he would look to the mind's insight into the proposition, "a square cannot be circular," he would see that it included the truism, "a square cannot be not square:" for incontrovertibly that which consists of curved lines is not square, and a circle is wholly curvilinear. Mill proclaims very loudly against Hamilton that what is self-contradictory cannot be sound philosophy: let him take his words home to himself.

Another example he borrows from a barrister, and it is to this effect. Two and two might make five; for example, it would do so in any region in which, when two and two things were put together, a fifth always "interloped." Really the argument seems childish, for the fifth object would never appear without a sufficient cause; and even though the inhabitants of the strange land never could discover what the cause was, at least they would rationally infer its existence, and never could form the judgment, "two and two make five." Yet Mr. Huxley has accepted the suggestion, and gravely told an American audience, "every candid thinker will admit that there may be a world in which two and two do not make four, and in which two straight lines enclose a space." If so, neither "candid" thought, nor any other kind of thought, has much intrinsic value.

From the same barrister Mill, whom Mr. Huxley follows obsequiously, shows how two straight lines may be judged to enclose a space. Writing lately in the *Nineteenth Century* against the Duke of Argyll, Mr. Huxley is inconsistent with his earlier view; for he lays it down "that omnipotence itself could not make a triangular circle." But let us go to the more original fount of wisdom, the barrister. "Imagine," says the learned counsel for the non-necessary truth of mathematical axioms,[12] "a man who has never had experience of straight lines through any sense whatever, suddenly placed upon a railway, stretching out in a straight line in each direction. He would see the rails, which had been the first straight lines he had ever seen, apparently meeting, or at least tending to meet, at the horizon. He would thus infer, in the absence of other experience, that they actually did enclose a space when produced far enough. Experience alone could undeceive him." Far more faults could be found with this piece of sophistry, which many grave writers patronize, than it is worth while to enumerate; suffice it to say, briefly, that in the supposed case a man, ignorant of perspective, erroneously judges from appearances two lines, which really are parallel, to be convergent: but he never judges that parallel lines can converge, for the notion parallel is nowhere shown to have entered his head. Here the barrister's random shot misses its mark utterly. No man, without secretly changing the meaning of his words, could intelligently say parallel lines, if prolonged, may meet. Even one of the empiricist school, Mr. Bain, has the wisdom to depart from his colleagues in this particular instance: "that two straight lines cannot enclose a space," he confesses, "is implicated in the very essence of straightness, as defined by mathematicians: to deny it would be a contradiction." It is against the convention, to which Mr. Huxley is a party, that terms should keep the same meaning.

[12] Quoted in *Mill's Examination*, ch. vi. p. 69.

The case of the barrister may be put in the form of question and answer. Q. "How may we reverse the apparently irreversible judgment, that parallel lines can never meet?" A. "By making a mistake, and fancying two lines to be convergent, which really are parallel." This is not a satisfactory conclusion. The view might have been given more speciously; but in its most specious form it would be dissolved by the words which Mill uses against Mansel: "I take my stand on the acknowledged principle of logic and morality, that when we mean different things," *e.g.*, parallel and convergent, " we have no right to call them by the same name."[13]

The result of an examination into Mill's conceived alteration, in what most people call necessary truths of mathematics, is to show the futility of his suggestions; and to convince us that there is no need to abandon the old views. Neither are we more inclined to believe Professor Clifford, in his solemn assurances, that while for the present our laws of geometry are, perhaps, only approximately true, for the future we cannot guarantee them to be even approximations. The necessity we continue to assert for geometric truths, we assert also for all other truth which shows itself to the mind to be evidently unalterable: it must be judged by the clear insight we have into the terms and their connexion, not by a fanciful theory, which derives all knowledge from the chance combination of sense-impressions, with the surmise that there is no assignable limit to the modes in which such com-

[13] *Examination*, c. vii. p. 101.

binations could be altered; that all judgment is the effect of association, and that all associations are possibly variable.

(c) Mr. Huxley's third sense given to necessary truth is that it signifies "facts of immediate consciousness"—"our sensations," he says elsewhere, "our pleasures and our pains, and the relations of these, make up the sum total of the elements of positive unquestionable knowledge." He does not exactly mean that there is no other knowledge: but that no other is beyond a question. Against the sufficiency of this view it has to be urged, that facts of consciousness are in themselves contingent, not necessary: and that what we regard as our chief necessary truths, though knowable to us only through facts of consciousness, are universal principles, not specially limited to facts of consciousness.

Moreover, facts of consciousness, as accounted for by the empiricist school, are made to appear in anything but the guise of necessary truths; rather they are reduced to a position of great confusion and uncertainty. Truism as it may appear to be, when we say "what we feel we feel," yet empiricism manages to obscure this act of self-consciousness. Mr. Bain, as we have seen, makes the matter one of a postulate for which no reason can be given. Mr. Spencer[14] declares that "a thing cannot at the same instant be both subject and object of thought," that "no man is conscious of

[14] *First Principles,* Part I. c. iii. § 20; *Pyschology,* Part II. c. i. § 59.

what he *is*, but only of what he *was* a moment before;" man is not conscious of his present, but only of his immediately past state; man holds in memory what he never held in immediate perception. In the same spirit M. Comte had written: "In order to observe your intellect you must pause from activity; yet it is this very activity you want to observe." If you cannot effect the pause, you cannot observe, and if you effect it, there is nothing to observe." Which words Dr. Maudsley[15] approves, and supports them by the principle that "to persist in one state of consciousness would be really to be unconscious: consciousness is awakened by the transition from one physical or mental state to another."

We have not yet arrived at the stage for discussing consciousness, but the passages quoted are to our point, because they show, that unsatisfactory as it is itself to take "necessary truth" to mean "facts of consciousness," the school of empiricists double that unsatisfactoriness by the difficulties they throw in the way of all consciousness. On this ground alone Mr. Huxley, if he were true to his authorities, as he need not be, would be disqualified from saying "we have seen clearly and distinctly, and in a manner which admits of no doubt, that all our knowledge is knowledge of states of consciousness." Yet this is his assertion: and it agrees with his

---

[15] *Physiology of the Mind*, c. i. Mill controverts Comte's views about Psychology. (*Logic*, Bk. VI. c. iv. § 2.) Of course Comte admits that somehow we do know our thoughts by reflexion. (*Philosophie Positive*, i. 35.) Mr. Huxley repudiates Comte's attack on self-introspection. (*Hume*, p. 52.)

third meaning of *necessary* truth, which, at best, is quite insufficient.

Three descriptions of necessary truth having been passed in review and found wanting, it remains that we argue in behalf of that fuller sense of necessity which undoubtedly is required, if man's position as a genuinely intelligent being is to be vindicated.

2. Our argument shall begin from admissions made by adversaries, who, when thrown off their guard, speak not according to the exigencies of a false theory about associated ideas, but according to the intellectual insight which is theirs by nature.

(*a*) If no truth can with certainty be shown to be more than a *de facto* association under present experience, it ought to be impossible to arrive at any element of *absolute* morality. Yet adversaries do make it a point of absolute morality that truth itself is, at all costs, to be held sacred. Whereas they ought always to say what Mr. Leslie Stephen says at least once,[16] namely, that "if in some planet lying were as essential to human welfare as truthfulness is in this world, falsehood may be there a cardinal virtue;" nevertheless they do say with Mr. Mill just the opposite, that it is better for human kind to suffer eternal misery than compromise the truth. The passage[17] is well known in which Mill declares, that rather than call any being good, who is not good in the human meaning of the word, he would go down for ever into Hell. Hereby he asserts a very strong conviction as to the absoluteness of

[16] *The Science of Ethics*, c. iv.   [17] *Examination*, c. vii. in fine.

moral truth, not only in this world but in the next, not only in man but in the Supreme Being. This is more than we could logically expect from a man who professed to doubt, whether a changed experience might not render inconceivable things now regarded as conceivable, and, on the other hand, render conceivable things now regarded as inconceivable; or, after Mill's own phraseology, dissociate the ideas of any present conceivability, and associate the ideas of any present inconceivability. If truth were indeed at its root, what mere empiricism makes it to be, it is impossible to show a valid reason why man should, in all cases, rather die than lie: and why Mr. Huxley can affirm "that the search after truth, and truth only, ennobles the searcher, and leaves no doubt that his life, at any rate, is worth living." Only when you give truth and goodness their foundation in some absolute necessary worth, are you able to show that between truth and untruth, right and wrong, the difference is as between Heaven and Hell. No wonder, then, Mr. Bain is puzzled, on his own principles, to justify a worship of truth for truth's sake, and has to apply the theory about means getting mistaken for ends.[18] "Associations," he pleads, "transfer the interest of an end of pursuit to the means. The regard for truth is, and ought to be, an all-powerful sentiment, from its being entwined in a thousand ways with the welfare of human society. We are not surprised if an element, of such importance as a means, should

[18] *Mental Science*, Bk. II. c. i. n. 34.

often be regarded as an absolute end to be pursued irrespective of consequences, whether near or remote." Nevertheless, a more correct insight occasionally asserts itself in the mind of the empiricist, and he becomes, in relation to his own dull principles, *splendide mendax*.

(*b*) But not only in the matter of morals, where it may be suggested that grandness of sentiment may have gained a momentary victory over clear thought, but even in the region of cold clear thought itself, adversaries are betrayed into admissions of metaphysical principles strictly so called. It is all very well to refuse attention to these admissions. Mr. Leslie Stephen, in answer to very forcible difficulties urged by Mr. Balfour, may reply with lordly disdain, as he has done in *Mind*, that he simply steps over metaphysical puzzles, and so reaches science; and he may own to only one exception: " To believe anything is the same as to disbelieve its contradictory: this is all the dogmatism to which I can plead guilty." Well, that one article only is fatal to empiricism, and has proved too much for Mill's powers of defence: besides, there are many other articles of which Mr. Stephen can be " proved guilty," even though he does not " plead guilty."

All that is needful is, to employ a means of conviction, which the late Dr. Ward used to employ with good effect.[19] He used to urge upon men of the school of Hume, that really, throughout their polemics, they were relying on the absoluteness of

[19] See the Preface to his *Philosophy of Theism*.

those very metaphysical principles, which they were labouring to prove only relative and contingent. To verify the force of this contention we have only to take up their books. It is not without an assumption of his own absolute knowledge that Comte can say, "There is only one absolute principle, namely, that there is nothing absolute."

Hume himself, in a sense which requires more sifting than can be afforded here, refuses to admit the validity of the inference, whereby, from past changes in nature, belief in the constancy of the same sequences for the future is derived. Why this refusal? Because he sees in the inference none of the demonstrative force that he acknowledges in the sciences of quantity and number, in which "reason is incapable of variation; the conclusions which it draws from considering one circle are the same which it would form upon surveying all the circles in the universe." On the other hand, empirical investigations are declared to want this invariability: "All inferences from experience, therefore, are effects of custom, not of reason."[20] He distinguishes a mathematical from a physical truth by saying that the former does not allow of any contradiction, whereas the latter might not be what *de facto* it is; and so far as facts are merely empirical, it is absurd to talk of them as demonstrable. He claims that his theory of causality upsets the common principle, that every event must have a cause, because upon this theory "we may easily conceive that there is no *absolute and metaphysical*

[20] *Inquiry*, Part I. sec. v.; cf. Part III. sec. xii.; Part I. sec. iv.

*necessity* that every beginning of existence should be attended with such an object."[21] Thus he requires for the establishment of a principle of human certitude, "absolute and metaphysical necessity," and rejects a most widely received axiom on the supposed defect of such necessity. Here is the tacit confession that every conclusion valid in reason must be drawn in virtue of some "absolute metaphysical necessity." Explicitly asked to make this confession, the empiricist would demur: implicitly, in the very act of using his reason, he yields his acknowledgment. He is constantly recurring to the phrases, "I see no necessary connexion," "I see no compelling evidence," "The conclusion is not inevitable,' and on these pleas he considers himself justified in stopping short at a probable assent.

It takes up too much space to transpose long quotations into these pages; but whoever wants a further illustration of how empiricists tacitly suppose metaphysical principles, need only read Mill's Preface to his *Logic*. There it will be seen how absolute is the character which Mill gives to logic; how carefully he submits all sciences, under pain of becoming unscientific, to the jurisdiction of the logician; how little he thinks of repudiating all necessity, or allowing for a possible alteration of experiences. Only two sentences shall be quoted, in which the noteworthy words shall be italicized. "Logic points out what relations *must* subsist between the data and *whatever can be* concluded from them: between the proof and *anything which*

[21] *Treatise*, Part III. sec. xiv.

*it can prove.* If there be any such *indispensable* relations, and if *those can be precisely determined, every particular branch of science,* as well as *every individual in* the guidance of his conduct, is *bound* to conform to these relations under penalty of making false inferences, which are *not grounded on the reality of things.*"

Of course it may be possible to trim these utterances into some sort of conformity with Mill's metaphysics; but the process is one of mere torture on a Procustean bed.

3. It remains that we ground certitude upon its only satisfactory basis of metaphysical principles, which have absolute necessity and universal validity. We can know metaphysical truths in the strict sense of the phrase.

A modern paradox is the denial by adversaries at once of necessity, of free will, and of chance. Hume[22] had led the way, saying, "Necessity is something that exists in the mind, not in objects." "Necessity," Mr. Huxley repeats, is but "a shadow of the mind's throwing," an "intruder" that he "anathematizes;" he claims to be a necessarian without being a fatalist, because he regards necessity as having only a logical existence. Free will he equally repudiates, and he would laugh at chance as a factor in scientific calculations. Necessity, free will, chance—these he does not recognize; but he adds, "Fact I know and law I know."

One point, at any rate, is asserted here; and while we cannot agree with Mr. Huxley's denials,

[22] *Treatise,* Part III. sec. xiv.

fortunately we can agree with his assertion of fact and law. We yield to none in putting fact and law at the foundation of all things, so far as God may be called (not indeed in the etymological sense) the first Great Fact, giving the law to all others. The substitution asked for in *Faust*, whereby " in the beginning was the Word," should give place to " in the beginning was the deed," has no point at all as directed against the reality of the Creator.

Next, what sort of a fact was this first fact? Not a chance fact, for that has no meaning: nor a free fact, for that is absurd in a first origin : but a necessary fact, for that alone will satisfy the requirements of sound reason. Necessity being thus at the root of all being, is therefore at the root of all truth ; the existence of the primal Being, its nature, its whole condition—this was the one great original necessity. Hume,[23] therefore, is too sweeping in his assertion, when he says, that of no fact is the contradictory inconceivable. It is inconceivable that the prime fact of existence should be reversible.

Here, therefore, is the foundation of metaphysical truth : here is "fact and law," but bound up with the anathematized " necessity." For the nature of necessary Being inevitably gives rise to certain necessary truths about being, on account of the identity between truth and being. But now observe, as a matter of great importance, that for the individual investigation it is not requisite, that before perceiving a truth to be of metaphysical necessity, he should have set before himself the origin of all

[23] *Inquiry*, Part III. sec. xii. ; cf. Part I. sec. iv. in initio.

things and of all truth, as in the sketch just given. It is enough that the intellect should clearly contemplate some of the easier first principles, and judge by evidence and insight. "The same thing cannot be and not be, at the same time and under the same aspect:" "nothing can begin to be without a sufficient reason for its commencement:" "things, equal to the same thing, are equal to one another." The simple understanding of these terms and of their interrelations is metaphysical certitude, necessary, universal, beyond all contingency. Evidence and insight—these are the things to insist upon, in opposition to the mere *de facto* experiences and associations, which Mill, at times, makes all in all. To set these latter in the place of supremacy is to yield to an utter scepticism, such as will presently be shown to be impossible. Mr. Huxley is fully aware into what an abyss the denial of insight into necessary objective truth, and the substitution of mere empiricism, inevitably conduct the speculator, who has logic and courage to follow his principles to their conclusions. Accepting Hume's principles, he boldly proclaims[24] that "for any demonstration which can be given to the contrary, the collection of perceptions which make up our consciousness may be only phantasmagoriæ generated by the *ego*, unfolding its successive scenes on the background of the abyss of nothingness."

Is the reader willing to go this length? If not, the only remedy is to keep a firm foothold on metaphysical certitude; for assuredly there is

[24] Huxley's *Hume*, c. iii. p. 81.

error in the supposition of Mr. Carveth Read, that "to doubt the possibility of necessary cognitions is not the same thing as to doubt the possibility of actual and objective cognitions." If there are no "necessary cognitions," that is, cognitions of necessary truth, then there is no fixed basis whereon to found the cognition of contingent facts or laws. Some support must be found for the contingent outside of contingency, that is, in necessity.

It is satisfactory to find a confirmation of the doctrine, that metaphysical truth is to be judged by evidence and insight, rather than on a theory of empirical associations, in the better utterances of Mill himself. Already we have seen that he asserts "belief" to be something different from association of ideas. If he had seen only this much, he had seen enough to warn him against judging the validity of the three great axioms of metaphysics—the principles of Identity, Contradiction, and Excluded Middle[25]—almost solely on the ground of conceivability as regulated by association. But Mill goes beyond the mere proposition that belief is more than association: for when speaking of evidence in relation to belief, he says:[26] "Inasmuch as the meaning of the word evidence is supposed to be something which, when laid before the mind, induces it to believe; to demand evidence when the belief is insured by the mind's own laws, is supposed to be appealing to the intellect against the intellect. But

[25] *Examination*, c. xxi.    [26] *Logic*, Bk. III. c. xxi. § 1.

this, I apprehend, is a misunderstanding of the nature of evidence. By evidence is not meant anything and everything which produces belief. There are many things which generate belief besides evidence. *A mere strong association of ideas* often causes a belief so intense, as to be unshakable by experience or argument. Evidence is not that which the mind does or must yield to, but that to which it ought to yield, namely, that by yielding to which its belief is kept in conformity to fact. To say that belief suffices for its own justification, is denying the existence of an outward standard, conformity of opinion to which constitutes its truth. A mere disposition to believe, even if supposed instinctive, is no guarantee for the truth of the thing believed." Agreeing with Mill that the mind must conform in its true beliefs to an outward standard, we have defended metaphysical truth on the ground that it has an outward standard in the objective evidence, which the mind perceives, and to which it conforms. But of evidence we must treat hereafter.

II. In passing from metaphysical to physical certitude, the transition is between two categories of Being, which Aristotle recognized under the names of necessary Being and contingent Being (τοῦ ἐξ ἀνάγκης ὑπάρχειν and τοῦ ἐνδέχεσθαι ὑπάρχειν). The ultimate possibilities of all things created are settled by metaphysical necessity, following inevitably, as is shown in Ontology, from the nature of the First Being and His powers of creation. Yet when the possibilities come to be actualized in the world, there belongs to them a lower order of necessity, which we call physical, and which, resting upon conditions that need not have been fulfilled, may be called contingent. Contingent necessity may seem a paradox, but it is easily explainable. Physical necessity rests upon a double contingency, on God's free election to create at all, and on His further free election of one out of many eligible plans of creation. The *de facto* elements, their number and original collocations, were matters of choice. But the system once established has intrinsic laws of action, which according to some theories of matter could not be altered without putting a different set of substances in place of the actually existent, while other theories would not so rigorously identify mode of action with substance. These laws we can partially detect, not by intuition or *a priori* argument, but by arguing back from effects to causes.

1. The sum total of created things and of their forces is regarded as a constant: so that we speak of physical nature as of a fixed aggregate, not liable to increase or diminution of parts. If it be asked how this fact can be known, the answer is, that our only natural means of discovery is by very wide observation. Undoubtedly God, if He had liked, could have put us into a world where He frequently took away old agencies and introduced new, or suddenly altered previous arrangements. Or He could have framed a world, different parts of which were composed of quite diverse elements, such even that no inter-action could go on between some parts and others. No one need have been very much surprised, had an old opinion proved to

be true, and had the heavenly bodies shown, that they rejected all kindredship with the physical constituents of our planet. Yet it would have been inconsistent with the essential Wisdom to have placed us in a creation, where the variability was so great, as to reduce us to absolute bewilderment, or to the position of dwellers in chaos, who could not familiarize themselves with their outer surroundings, or so accommodate themselves to their circumstances as to be able to continue the life of the race. There must then be some uniformity of nature, and it becomes urgent upon us to distinguish different uses of that phrase.

(*a*) The most radical meaning of all, is that like agencies, under like circumstances, will always have like effects. Messrs. Bain and Pollock, not admitting the principle of efficient causality, have agreed in maintaining, that for anything we can know to the contrary, the mere lapse of time may make an alteration. On this point Lewes rightly took the other side, and held, though in an imperfect manner, that the circumstance of time, as such, is irrelevant, and that the principle is an *a priori* truth. Time, as time, never alters anything; but alteration is due to the activities, which, in time, produce their effects. What is relevant as regards time is this: created things continue in their communicated existence only by virtue of the constantly supplied support of Him who originally gave them being: and on this score, a natural object has no intrinsic power of prolonging its own duration. But when we speak of like agencies having like effects, the presupposition is that they are preserved in their proper natures; else we could not call them like. The non-theistic school of philosophers will not approve of the mention of creation and conservation; but they must remember that questions of this sort necessarily drive us back into the theory of first origins; and that those who simply have no view as to the beginning of things, or as to the production of existant objects, must allow that they have a great and fatal deficit in their philosophy.

This something which is wanting shows itself in many curious opinions about a means of origination, which ultimately may be reduced to the illogical idea of chance. As theism is true, no apology is needed for using it to settle points, which otherwise cannot rationally be discussed: and we must consider the agnostic position as quite unfitted to give its occupiers the safety, which they vainly imagine that they possess in the word, *ignoramus*. On the plea that they do not know anything to the contrary, they speak of it as a possibility, that there might be a world where things spring into, and out of existence, as it were spontaneously and capriciously: in which case, as Professor Clifford suggested, it would be worth while trying to settle what objects were given to such vagaries; whether, for instance, buttons were prone to these pranks. The great mystery, what becomes of all the old pins, might be more hopefully investigated on the hypothesis of sudden ceasings to be. Wild as the notion may seem, it is contained in Mr. Bain's[27] solemn

[27] *Mental Science*, Bk. II. c. vi. n. 9.

announcement: "That every event must be preceded by some other event is obviously not necessary in the sense of implication, and the opposite is not self-contradictory. There is nothing to prevent us from conceiving an isolated event. Any difficulty that we might have in conceiving something to arise out of nothing, is due to our experience being all the other way. If it were not for habit there could be no serious obstacle to our conceiving the opposite state of things to every event being chained to some other event." Thus to abolish the principle of efficient causality is to take away all genuine science; for in that case there could be no proof that uniformities would continue, not even, strictly, that they had existed in the past. To guard against this chaotic result, we state the first sense of nature's uniformity to be the *a priori* self-evident principle, that from like causes, under like circumstances, uniformly constant results may be relied upon to follow.

(*b*) The second sense of uniformity in nature is *a posteriori*, as the first was *a priori*. The first says, like agencies, under like conditions, will always have like effects; the second says, the sum total of physical agencies in the world is constant, neither matter nor its inherent forces suffer increase or decrease. This is not going as far as the Law of Conservation of Energy; but it is its foundation. The asserted uniformity cannot be verified in every separate detail; but it is what all observation of nature goes to establish.

(*c*) The third sense of uniformity is again a matter of observation. It is noticeable that in some climates, for instance, the dry and the rainy seasons are calculable almost to a nicety: whereas here in England, which has according to an American authority, "no climate but only specimens of all sorts of weather," we take it as a matter of no surprise that fair or wet weather should predominate in any of the four seasons. The laws are fixed for us, as for the most regular of climates; but whereas, for the latter, they result in obvious regularity, for us they result in apparent irregularity. Speaking of the uniformity of nature from this point of view, we have evidences of it in many recurrent phenomena, such as day and night, the seasons of the year, planetary conjunctions, secular variations like those effected by the precession of equinoxes, and lastly successive stages of animal life in one and the same individual. Thus the universe on which we dwell, in many of its phenomena, does not, but in many also does, present us with detectable periodicities; and these we may fairly call *uniformities of Nature*.

But another physical universe is possible, where such recurrences would be so rare as to give to an observer, having the average life of man, no token of regularity. Uniformity would be there in the first sense of the word and in the second; matter and force would be constant, physical causes would keep rigorously to their laws; still the combinations would be so various as to present an appearance of chaos. Elementary laws would result in complicated effects, without discernible law of complication. Compared with such a possible world, ours we call uniform, because of its many observed recurrences.

2. If we hold by the several truths just enunciated, we shall be saved from the sad lot of empiricists, who have to take refuge in "a primitive instinct," or in an "unaccountable adaptation of our beliefs by the Creator of our faculties," in order to explain, why it is that we rely on our past experiences for knowing what nature will do in the future.[28] Our reliance is rationally grounded on the three uniformities above described; one *a priori* and quite necessary, the other two *a posteriori* and necessary only inasmuch as God cannot fail to give to His works their strict requisites for the purpose they are meant to serve. This is theism if you like, introduced into philosophy; but theism is itself philosophic, and so necessary to philosophy, that if you deny it, you have no stable basis for physical truth, but at best a hope, logically quite unjustifiable, that the course of things will go on with that orderliness, which hitherto you have known it to observe. Further than this the non-theist cannot advance: for him any time there may be "chaos come again." Mill[29] is quite open in his avowal that on his principles, there may be a planet where "events succeed one another at random, without any fixed law," and that " it is perfectly possible to imagine the present order of the universe brought

---

[28] Examples from one who so speaks have already been given under the head of Metaphysical truth, for reasons there stated. See the present chapter under the headings I. 1.

[29] *Logic*, Bk. III. c. xxi. § 1.

to an end, and a chaos succeeding, where there is no fixed succession of events, and the past gives no assurance of the future." In the same spirit and on the same principles Mr. Huxley writes in his *American Lectures*: "Though we are quite certain about the constancy of nature at present, it by no means follows that we are justified in expanding this generalization into the past, and in denying absolutely that there may have been a time when events did not follow a fixed order, when the relations of cause and effect were not fixed and definite, and when external agencies interfered in the general course of nature." There are statements here fatal to physical science, which can be preserved from extinction only by holding on to principles we are advocating, not indeed as anything new, but as the common possession of unsophisticated mankind.

3. Wishing now to maintain the power of the human mind to reach physical certitude, we much need a distinction between two classes of efforts—those more ambitious efforts which often do not get beyond probability, and those humbler efforts which often reach full assurance. Against the absolute certainty of the sun's rising to-morrow it may be urged, that even though our system were clearly explained as to its planetary movements, still there would remain elements of doubt. For instance, we are told that the whole system is travelling in space; that the stars are closing up behind our course and opening out before; and that it is not quite sure that we shall not come suddenly under perturbing influ-

ences as yet unsuspected. It is admitted that the danger is a *minimum*, as far as we can calculate: but nevertheless there is a particle of undispelled doubt, nay some would say far more than a particle. Well, give this theoretical doubt its due, and, after all that astronomers and even theologians who speak of providence, can bring forward to comfort the timid, suppose it to remain undissipated. The sun's movements are not the easiest of our physical inquiries, and it is precisely in our more complicated or our abstruser questions, as for instance whether the law of the inverse squares applies to gravity at minutest distances, that we may allow some truth to Mr. Huxley's declaration, "that our widest and safest generalizations are simply statements of the highest degree of probability."

But take the simpler case of letting a stone drop to the earth. Arrange your own circumstances, break off a piece of sandstone from a quarry, which you know well; get out of the way of all scientific apparatus, on to the open plain, and there, relaxing your hold upon the stone, leave it to nature's forces. You may not know all about gravity; there may be many forces acting on the stone about which you are ignorant: still you have physical certainty that the stone will not stand in mid air. As to the possible unknown forces, you have sufficient experience to warrant the conclusion about what they will *not* do—that they will *not* arrest the fall to the ground. It is a physical certitude of this simple nature that we often want for purposes of daily life, and sometimes for such a religious purpose as verifying a miracle. Unless he had in mind the grade of certitude, about which I spoke before, and of which his example would give a good illustration, it is hard to see what De Morgan, in his *Logic*, can have wanted to show, when he wrote:[30] "I know that a stone *will* fall to the ground when I let it go, and I know that a square number must (in a given case) be equal to the sum of odd numbers: and though when I think, I become sensible of more assurance for the second than for the first, yet it is only on reflexion that I can distinguish the certainty from what comes so near to it." Is not this only another case of playing fast and loose with the word certainty? "I know that the stone *will* fall:" and yet the knowledge is only what "comes near to certainty," but is distinguished from it. We should say that the certainty from which it is distinguished is not certainty in general, but that special sort of certitude which carries with it *must* instead of *will* or *is;* or that one is metaphysical, the other physical certainty. But both are full certainties.

4. There still remains the objection, what about miracles? If God can interfere at any moment with the course of nature, how determine in any case that He does not interfere? In reply we must say

---

[30] De Morgan gives us expressly his views on the grades of certitude. (*Logic*, chap. ix. in initio, p. 171.) Speaking of the knowledge we have of our own existence, and that two and two make four, he says: "This absolute and unassailable feeling we call *certainty*. We have lower grades of knowledge, which we usually call *degrees of belief*, but they are really *degrees of knowledge*," *e.g.*, man's belief that yesterday he was certain about two and two making four.

that the objection is not insuperable: in many instances we may be sure there is no miraculous interposition. For God has sufficiently shown us, by experience and by reasons of fitness, that miracles do not come in capriciously, so as to make the whole of life a puzzle to us: but they are wrought only occasionally and for proportionate ends.

> Nec Deus intersit, nisi dignus vindice nodus
> Inciderit.

Surely there are trivial circumstances in our lives, where we can see that there is no adequate occasion for miracle, and where, in consequence, we may know that none will be performed. And as for Descartes' fear of a mischievous demon, who may be always tricking us, it belongs to God's providence to hold in check the limited powers which even the evil spirits, by natural endowment, possess.

Some may object to Divine providence as a factor introduced into philosophical considerations. But a factor it is in the world's physical course, and as St. Augustine long ago pointed out, if we neglect this factor, then *actum est de philosophia*. Those, however, who exaggerate the possibility of Divine interference seem not at all to realize what they are committed to, when, because of it, they have taken up the position, that never can we be quite certain of a physical fact or sequence. They fail to observe that they cannot at once hold this position, and at the same time claim to be sure that there are, or have been, a city of London, a man called Napoleon, and a plague known as the Black Death. When they speak of miracles as always possible, they forget all the ridiculous interferences, which, on their theory, it is not incredible that God may work; for if no physical event is safe from the suspicion of miracle, then it is not certain that to-morrow all men will not be walking on their hands, all corn will not become poison, and all sand will not turn into gold. Really with the fullest allowance for large possibilities in the way of unsuspected miracles and for the inadequacy of our knowledge about any one of nature's ultimate laws, still we must not go the length of conceding our complete inability, to be certain of physical truths, past and present. As to the future, if any one likes to fancy an instantaneous arrival of the end of the world, it would be difficult to plead anything against him, except from the signs given in Scripture about what is to precede the consummation of all things terrestrial, and from the fact, that the immediate future of our universe is, to some degree, calculable from its known present. Conjectures are even made about the natural causes of a final period to be put to the order which now prevails.

### ADDENDA.

(1) That the exaggerated manner in which some urge the association theory, leads to the denial of all immutable truth, cannot but be known to any one at all acquainted with our English writers on philosophy. To take a single specimen, we have Dr. Maudsley[1] telling us to give up as hopeless "infinite, absolute

[1] *Physiology of the Mind*, c. v. p. 141. (2nd Ed.) Compare Hume, *Treatise*, Part III. sec. xii.

truth." If he means only that we cannot grasp truth in all its infinity, he is obviously right; but he means more and worse. He says, "Because each one has a certain specific nature as a human being, and because the external nature, in relation with which each one exists, is the same: therefore are inevitably formed certain general associations which cannot without great difficulty, or anywise, be dissociated. Such are what have been described as the general laws of association, in which all men agree—those of cause and effect, of contiguity in time and space, of resemblance, of contrast; in all which ways, it is true, one idea may follow another, though also probably in other ways. The universality which is supposed to belong to the ideas of cause and effect, of the uniformity of nature, of time and space, has been supposed to betray an origin beyond experience," that is, beyond mere empirical association. "Nevertheless, it is hard to conceive how men, formed and placed as they are, could have failed to acquire them, *and still more difficult to conceive, how they could even have been supposed to have any meaning outside human experience, to have an absolute, not a relative truth.*" Thus the law of causality is true for men, with a mere relative truth, and has no absolute value for all intelligence; a theory which robs science of all its glory, and is made worse by what follows. "The belief in the uniformity of the laws of nature is a belief which is developed of necessity in the mind, in accordance with the laws of the nature, of which mind is a part and product. The uniformity of nature becomes conscious of itself, so to speak, in the mind of man: for in man, a part of nature and developing according to nature's laws, nature attains to self-consciousness. To declare that a theory is conceivable, is to declare that conception has limits based upon experience, not to limit the possibilities of nature." All thought thus becomes a sort of *de facto* pattern, worked out in the mind of man by his surroundings: whilst other surroundings would have worked out quite a different pattern, and no pattern has any absolute value. What is true of mere sensations is thus extended to the highest acts of intellect. Hence no fixed system of philosophy is possible; at best we can but have ideas suitable to our own age and *Zeit-Geist*, or spirit of the time. As Mr. Pollock[2] puts it, "Science makes it plainer, day by day, that there is no such thing as a fixed equilibrium, either in the world without or in the world within: so it becomes plain that the genuine and durable triumphs of philosophy are not in systems but in ideas." But what is the value of ideas, which condemn each other by refusing to fit into consistent system? Let us take the instance of a few "ideas," which have been framed to represent our condition as regards the knowledge of nature.

(2) Reid[3] has told us, far more piously than wisely, "God hath implanted in our mind an original principle by which we believe the continuance of the course of nature, and of those connexions which we have observed in the past. Antecedent to all reason we have an anticipation that there is a fixed and steady course of nature." Brown,[4] in default of a belief in real causality, is also obliged to fall back on Providence, appealing to "the instinctive tendency wherewith God has endowed us in view of the circumstances in which we are placed." Mr. Bain[5] leaves out all mention of a bountiful Provider,

[2] See his *Life of Spinoza*, in fine, p. 408.
[3] *Human Mind*, c. vi. sec. xxiv. p. 198.
[4] *Inquiry into the Relation of Cause and Effect*, Part III. sec. v. p. 249.
[5] *Logic*, Appendix D. p. 273.

whose existence he would consider unverifiable, and points simply to blind tendency. He asserts that there is "a primitive credulity, which every uncontradicted experience has on its side," "an initial believing impulse of the mind, which errs on the side of excess, and which, if nothing has happened to check it in a particular case, will be found strong enough for anything." Neither Mr. Bain's theory, nor any philosophy of Hume's school, will give to physical science a rational basis: and this is a serious consideration for those who may feel tempted to grasp at the simplicity of experience and association, when put forward as explanations of well-nigh everything that can be rationally explained.

(3) With metaphysical and physical truth alike overthrown, with the very principle of contradiction undermined, it is no wonder that we have philosophies in which contradictions abound.

Nor can the work of clearly pointing out these contradictions, be looked upon as a useless sort of criticism. Take the case of Mill for instance. Mr. Jevons, disgusted with the task of having to teach his system for several years, entered a protest by publishing a list of the inconsistencies which he had come across, many of which are undoubtedly to be found in the author. This is a most legitimate and effective way to discredit a philosophically discreditable writer, and serves the very good purpose of doing something to check the spread of ruinous principles. It is, then, somewhat difficult to see the force of the objections made by the Editor of *Mind*, when he says that Mill's inconsistencies are known; that no one is exactly a follower of Mill; and that those who admire him most and owe him most, take leave to dissent from him when they think good. All this may be true: and yet, since Mill has given to Hume's philosophy about as fair an appearance as any other author has succeeded in imparting to it, the labour is a worthy one, to show in detail the essentially contradictory character of a bad system. A list of Mill's inconsequences and contradictions should be kept as permanently on the book shelves as his own works—the antidote ever by the side of the poison. Perhaps it was because he rose up among a people who had long neglected philosophy, and whom he helped to rouse into inquisitiveness on the subject, that Mill's undoubted cleverness met with so much success in the propagation of irrational principles. But there is no reason why Englishmen should go on worshipping the god of unreason: especially when they remember Mill's wretched education from earliest years. He is always to be spoken of more in pity than in anger; but when we read Mr. J. Morley's extravagant praises of him, and profuse acknowledgments of indebtedness to him as a teacher, while we understand better Mr. Morley's position, we also understand the need of having the hollowness of the teacher sounded and made known to all.[6]

[6] See two articles on Mill in Mr. J. Morley's *Miscellanies*.

# CHAPTER VI.

## THE ORDER OF PRECEDENCE BETWEEN NATURAL AND PHILOSOPHIC CERTITUDE.

*Synopsis.*
1. As a fact, non-philosophic or natural knowledge has preceded philosophic.
2. What is meant by philosophy in general.
3. Applied Logic is a part of philosophy.
4. The justification of one who, without mastering scientific logic, cultivates the other sciences.
5. How scientific arises out of non-scientific logic.
6. Consequent deduction of practical principles, whereby to judge and choose a system of philosophic certitude.
7. Hopeless search after a philosophy of certitude, built up step by step like Euclid's geometry, and never anticipating the results of a future step.
8. Parallel case of trying to arrange the sciences hierarchically, or in order of subordination.
9. Short maxims summarizing the practical results of the chapter, and warning the reader against the extravagances of philosophizing.

1. WE must next begin to handle the question, about our real possession of certitude concerning things. All along the affirmative answer has been tacitly assumed, as it must be assumed by whoever professes to be conducting a rational discussion: but it is now time to talk explicitly about the subject. Philosophy, though an inevitable development of mental culture, belongs rather to the *bene esse* than

to the *esse* of intellectual life. If ever luxuries precede necessaries, as in the priority of metrical over prose literature, there is some accidental reason for this apparent inversion of right order. The early Greek philosophers found verse decidedly an easier way of giving currency to their opinions: so that when Heraclitus of Ephesus made the experiment of trying to invent a prose style that should have scientific accuracy, he brought down upon himself, perhaps not solely because he wrote in prose, the epithet of ὁ σκοτεινός, "the Obscure." But before any systematic philosophy, which is worth the name, and is not a mere fantastic cosmogony or something of that sort, there must go a fair development of the intellect, by its working, we do not say unphilosophically, but non-philosophically.

2. By philosophy is here meant "the knowledge of things through their ultimate causes."[1] All science agrees in being *scientia rerum per causas*, where the word "cause" is used in a wide sense, to signify the *rationale* of things: but it is special to philosophy to investigate the very ultimate reasons of things. Not all parts of philosophy, as is plain, can be about things equally ultimate; but all parts are deservedly classed as ultimate investigations.

3. The subject of the present treatise is undoubtedly, in its own order, an ultimate inquiry: for it discusses the very radical question, What is the validity of human knowledge? The special sciences assume this validity, and upon the assumption observe, analyze, synthesize, and methodize. Applied

[1] "Scientia rerum per causas ultimas."

Logic has to take up the previous question, What is the guarantee of objective validity in observation, analysis, synthesis, and method ? Sometimes the man of special science laughs at the logician : but he would not laugh if he remembered that, unless Logic is valid, his own conclusions are of no scientific value.

4. And yet the man of concrete science need not be a philosopher, which is not the same thing as saying philosophy need not be true; without philosophy he is quite right to take the validity of his faculties, and his way of using them, for established. We may go some way with Balmez when he says in his *Fundamental Philosophy* : " If any part of science ought to be regarded as purely speculative, it is undoubtedly the part which concerns certainty."[2] For consider how we teach philosophy. We let a boy go all through his school course, which includes various sciences, but we do not ask him to study philosophy strictly so called. If he intends to take up this branch, we are glad of his deferring it for a few years more; and if he enters upon his course at the age of twenty-one, we are rather satisfied than sorry at the delay, because he brings to his task a maturity of years, which is usually indispensable for real philosophizing, as distinguished from learning systems by rote, or from learning how to manipulate stock phrases.

Here, then, we show our firm belief that stores of real knowledge, and even of scientific knowledge, may be gathered by the mind that has never turned introspectively upon itself to systematize its own laws. What we call natural knowledge we hold as quite valid: the mind observes, reasons, and reflects, and in the exercise of these faculties perceives its own powers, and is convinced that it acts rightly. At the same time there spontaneously occur these self-questionings, which, when systematized and answered, form a body of philosophic doctrine.

5. Philosophic logic, therefore, is natural knowledge rendering reflexly to itself an account of itself. Wonderful and most necessary to true intelligence is that power, whereby the mind can make its own thoughts the object of further thought: and herein lies one of the manifest discriminations of man from lower animals, and one of the proofs for the spirituality of the soul. We have not two intellects, the one ordinary, the other extraordinary; the one direct, the other reflex; but we have a single intellect to think, and to analyze thought, to do our common-sense thinking and our philosophical thinking.

6. Whence follows a golden rule—distrust that philosophy which is at utter variance with common sense. What Mr. Bain says apologetically for idealism, forms really the strongest presumption against it, namely, that language, as we now have it, is based on the contrary hypothesis, and so will not serve the purposes of the idealist. Mill,[2] too, is uttering his own condemnation, when he pleads unfairness in language; and says that if his theory of mind appears more incomprehensible than its rival, the reason is "because the whole of human

[2] *Examination*, c. xii. p. 213. (2nd Ed.)

language is accommodated to the latter, and is so incongruous with the former, that it cannot be expressed in any terms which do not deny its truth." It was one of Ferrier's pet declarations that "philosophy exists for the purpose of correcting, not for the purpose of confirming, the deliverances of ordinary thinking." If he had meant no more than that philosophy, like any other science, should correct some popular delusions, there would have been nothing against which to object; but he meant a substantial correction of ordinary thinking, and that he was wrong, his own untenable idealism is sufficient token. Hegel, too, was wrong, as his system again proves, when he asserted that "the mystics alone are fit for philosophizing." In another direction M. Ribot goes astray in his remark that philosophy has the value of mental gymnastics, exercising the faculties upon problems hopelessly beyond their grasp, and for that very reason calling forth the utmost efforts of the mind: just as a man might jump at a stretched string which he had no prospect of ever reaching, even with head or hands. Rather we should hold that, as the perfect Greek athlete was a man with flesh-and-blood muscles, trained to the utmost, but still of flesh-and-blood; so the perfect philosopher is a common-sense man, who has bestowed uncommon care on the scientific examination of his common sense, but only by the aid of that which he has been examining. A philosophy written from this stand-point will read as if written in the open air, not in some sickly closet, where body and mind have their natural health destroyed.

On the principle here maintained, philosophy must never do anything that is dead against natural reason, as, for instance, give it the lie direct, or doubt its evident convictions. More will be said of Descartes hereafter, but he is too apt an illustration not to be used at present. He professed to be able "seriously and for good reasons" to doubt such self-evident truths as the capability of his own faculties to acquire knowledge and the plainest axioms in mathematics. Now this was sawing off the branch on which he sat, and brought him to the ground, shattered beyond the possibility of rising again. It was philosophic suicide. Even Hume noticed that "the Cartesian doubt, were it ever possible, as it plainly is not, would be entirely incurable; and no reasoning could ever bring us to a state of assurance on any subject." Aware that it cannot create an intellect of its own, or discover an intellect that has not first spontaneously manifested itself, the scholastic philosophy accepts the position and makes the best of it, which best is not bad. It does not aim at a new kind of knowledge, a Soufi ecstasy or Hegelian dialectic, but only at elevating the vulgar knowledge, extending its range, and especially training it, by the aid of its own lights, to see its own highest principles of activity.

Hence the theory of knowledge, as proposed by the scholastics, whatever may be said of some details, at least in its essential parts has nothing that makes a heavy demand on the credence of the ordinary mind—such a demand, for instance, as is made by our pure empiricists, and our so-called

Neo-Kantians, who scarce have the first requisite of intelligibility, and who, so far as they are intelligible, are often extravagant. Indeed, the scholastic account so falls in with the view of the ordinary thinker, that the latter, when he takes up our treatises, is apt to exclaim: Is this what you call philosophy? Why, it seems to me that it needs no philosopher to point out that intelligence is intelligent; that what is evident is true; that the final test of understanding is, on one side, the actual experience of being able to understand; and other such plain propositions into which I can resolve your rather more elevated utterances. There is truth in these remarks, and a truth not to be disguised, nor shamefacedly admitted, but manfully recognized. Our philosophy does start from common sense, and can never shake itself free from its humble beginnings. It is a *terræ filius* by origin; but at least it is the offspring of a healthy soil; and now that it has dressed itself up and made the best of itself, it presents no ignoble appearance. Neither was its parent, natural knowledge, mere blind instinct; it had the same means at command as philosophy has, but its skill in the use of them was somewhat inferior: though it saw its way as it went, it had not the cleverness actually to draw a map of the course. Now it can not only make journeys, but write an account of them, and gives sketches by the way.

7. The nature of philosophy being thus explained, it is clear that we can never find what some seem to insist upon, and what Ferrier tried to give in

his *Institutes of Metaphysics*, namely, a philosophy of certitude built up after the plan of Euclid's geometry. Euclid begins with axioms, postulates, and definitions, and then he so piles proposition on proposition as never to need the conclusion of a later proposition as part of his proof of an earlier. But Euclid assumed those truths, which the philosophy of certitude has to discuss: what he had to prove lay all within the narrow department of quantity in extension, as represented by lines and angles. On the other hand, he who draws out the philosophy of certitude has to discuss the very faculties and principles which he must be using all the time, and cannot proceed a step without tacitly assuming the conclusions of pretty nearly his whole treatise.

Write any first chapter you like to your Book on Certitude, and see how far it is from involving only one simple idea or principle: see how much it already implies, upon which you will have afterwards to raise questions. You are going in general to ask if man can have real knowledge: and how can you help supposing all the time that he can? Relying on the veracity of the senses, in spite of its being so hotly canvassed a point, you refer to the writings of other authors, and in return you have recourse to the printed characters, which are to convey your thoughts to the world.

The reader, therefore, must be patient; and wait till he arrives at the end of the book, before settling, in his own mind, that the author leaves necessary matters undiscussed; and he must not expect a Euclidean inverted pyramid—a system rising, as it were, from a point and broadening as it ascends—to be erected where that style of structure is neither needful nor possible. He must not too readily take it for granted that there is illegitimate arguing in a circle, if he is referred about from chapter to chapter, or told to put off, till a subsequent chapter be reached, his search for various pieces of information. If it is better to refrain from plainly saying that many of our propositions cannot strictly be proved, it is not because this declaration would not contain a truth; indeed, it is eminently true; but because it is pretty sure, in nine cases out of ten, to be taken in the very false sense, that no satisfactory account can ultimately be given of the judgments we hold by, and that we can take our so-called knowledge only on blind trust. From such a view we must strongly dissent; and if some propositions in this treatise are called not strictly demonstrable, the meaning is that they are immediately evident, and do not admit of resolution into simpler propositions.

8. What has been said of the *Allzusammenheit*, "altogetherness," or interfusion of parts, in the philosophy of certitude, which forbids the orderly march of propositions that we see in Euclid, may be paralleled by the impossibility of putting the several sciences into exact hierarchical order. One objection which Mr. Spencer urges against Comte's classification, namely, that some of the earlier sciences have to wait for advances to be made in

the later, will always remain, whatever be the arrangement in way of subordination: and a quite perfect gradation is impossible. This is a fact, but it need create no great discomfort.

9. After having explained some wrong and some right conceptions as to the nature of philosophy, and having in mind the sad extravagances which the history of philosophy reveals in far too large a proportion of its pages, we may now draw a practical conclusion as to the sane method of philosophizing. We observe that the strain after the very knowledge of knowledge and wisdom of wisdom, has led to the neglect of the Apostolic precept, "Not to be wiser than it behoves us to be wise, but to think soberly."[3] Hence are suggested golden mottoes like these: "Moderation is the best";[4] "Be not wise beyond thy wits"; "Be not wise after the manner of the wiseacre"; "Philosophize not unto foolishness"; "Do not for the sake of philosophizing destroy the foundations of philosophy."[5] These and the like maxims the philosopher should keep in his mind as ballast, or else the mental balloon may quickly be found outside the element wherein man can breathe. With regard to how many a writer, Hegel, say, or Hartmann, or one of the old Gnostic evolvers of Æons, have we sorrowfully to exclaim: "Alas, poor man, he has taken the headlong plunge into the great inane: it

---

[3] "Non plus sapere quam oportet sapere, sed sapere ad sobrietatem." (Romans xii. 3.)
[4] μηδὲν ἄγαν, μέτρον ἄριστον.
[5] "Noli propter philosophiam, philosophandi perdere causas."

is hopeless trying to follow him, and he himself will never re-emerge!" The greater his powers, the more desperate, perhaps, is his condition; for, as St. Augustine observes, *Magna magnorum deliramenta doctorum;* or, as Balmez puts it, "There are errors which lie out of the reach of an ordinary mind"— words which for present purposes it may be allowable to understand so that they form a repetition of the *dictum* of St. Augustine. One thing this volume does promise the reader, that in it he shall never be asked to believe what to the plain Christian man is startling, or appeals to no intelligent principle within him. It has no propositions brought down from the region of the marvellous. Mr. M. Arnold has lately told us, that there has at length dawned in England a day for which, years ago, he could only hope; and that now it *is* here regarded as an objection to a thing that it is absurd. If ever such a day dawns for philosophy, how will its light dissolve the hazy reputation of many a once cherished philosopher!

# CHAPTER VII.

THE CHARGE OF DISCORD (OR AT LEAST OF WANT OF CO-OPERATION) BETWEEN NATURAL AND PHILOSOPHIC CERTITUDE.

*Synopsis.*
1. The asserted antagonism of Philosophy to Natural Certitude. (*a*) A thorough-going antagonism. (*b*) A partial antagonism.
2. The asserted want of co-operation between spontaneous and systematic thought, or between natural and scientific reasoning, can be explained by the consideration of certain facts. (*a*) When theory is not yet as wide as all the conditions of a problem, it is no disrespect to theory to supplement it by rule of thumb: theory co-operates to the extent of its powers, and there stops short. (*b*) By long habit the mind abridges its processes, and does not always follow out every logical step in an inference. (*c*) The spontaneous processes of the mind may very well be more successful than the reflex on many occasions.
3. Limits within which the doctrine in the chapter is to be taken.

THE philosopher's prying into his own mind has been compared to Aladdin's prying into his wonderful lamp; before, it lighted him to the most marvellous discoveries, afterwards, it became unserviceable. This accusation is urged by different authors in varying extent; with some the charge is one of downright antagonism between philosophy and natural certitude, with others it is one of want of co-operation or of harmony.

1. The assertors of antagonism must be subdivided into those who represent the opposition to be complete, and those who represent it to be partial.

(*a*) That philosophy utterly discredits the validity of ordinary reasoning is what we should gather from some of the stronger expressions used by Jouffroy. For example, he declares that reason "absolutely affirms human belief to be without a motive; it is by instinct that a man believes, and by reason that he doubts. When reason reflects upon its own work, scepticism is the inevitable result." This is but a repetition of Bayle, who declared that reason can not bear to turn her own light upon herself; that philosophic reflexion undoes all the mind's previous work and makes her a Penelope, unweaving at night what she had woven by day.

Against so blank a scepticism, as resulting from a philosophic examination of man's position in regard to knowledge, it will be the business of the next chapter to contend; so at present we may pass on to the milder subdivision of the first impeachment.

(*b*) At any rate, it is argued, philosophy is only in partial agreement with common certitude, and there is a partial disagreement. Speaking of the sceptic doubts which philosophy can throw on scientific principles, and of the practical progress of science in spite of these apparently demonstrated difficulties, Mr. A. Sidgwick thinks we must acquiesce in a certain disregard of what seems philosophically valid argument. "In the presence of all the acts of useful self-deception, which helps to make the world go round, may we not admit that theory and practice cannot as yet be safely presumed to coincide?"

A writer who has done good service to Catholic philosophy in this country, Dr. Ward, has more than once expressed an opinion which bears on the present discussion. Though a great stickler for logic, yet it was his deliberate view, "that there are several truths of vital importance, which are reasonably accepted as certain only on implied grounds of assurance, which have not as yet been scientifically analyzed; nay, of which, perhaps, the scientific analysis transcends the power of the human soul."

Out of the two authors quoted, we may frame a sort of common objection in this shape: Practical logic, as it may be called, outstrips the school logic, sometimes bidding us go safely forward, where the latter posts up a decided "No Road." Thus, at least, there is occasional opposition.

In reply, let us begin by distinguishing between what one individual and what another individual can do: as also between what any unaided individual may accomplish and what the collective force of human intellect may accomplish. The individual unaccustomed to the analysis of his thoughts may often have a genuine certitude, for which, nevertheless, he is unable to render a philosophic account, but for which another individual, trained in philosophy, would furnish a sufficient analysis. Next, beyond the individual, we have to take into account the accumulated labours of the race, especially of its ablest members working in conjunction upon the chief problems which present themselves for human investigation. What now are we to say of a professed certitude, which both the individual man and collective humanity have failed to support by producible motives? The certitude is, by supposition, merely a natural act: yet nowhere among men can immediate or mediate evidences be brought forward adequate to its defence. It has to be accepted on a general feeling that it is right; but how or why it is right, no one can exactly declare. Where is the instance of a certitude about a "vital truth" in this predicament? If such there be, about the only rational ground on which it could be defended would be by saying, that the race of men being rational, such a common consent could not have been produced except by some rational motives, however inscrutable some of these might be. But we may doubt whether any human certitude is so circumstanced. It seems more correct to maintain, that for every certitude, which is not self-evident, there is a producible analysis of motives. A perfect analysis may not be forthcoming, but at least a sufficient explanation may be offered. If the truth is self-evident, the self-evidence is the motive of belief; otherwise there must be some inferential evidence. At any rate, for a real certitude of the natural order, there must always be producible evidence.

By far the most pertinent reply to alleged instances of the difficulty we are now considering is to point out that the case is not one of full certitude, but only one of high probability, quite sufficient to act upon. We have no fear that the sun will not rise to-morrow; yet those items which are wanting to the full logical proof of coming events are just what cause our legitimate assent to fall a little below absolute. If the sun did not rise to-morrow, we should be ready to confess "Well, after all we had not absolute demonstration." Thus, as a fact, valid assent is not in excess of the premises, and practical logic does not really carry the intelligence further than speculative logic would allow. In all cases genuine certitude is strictly proportionate to its known motives.

2. Without being opposite, paths may not coincide; and when opposition, between the ways along which spontaneous reasoning and philosophy respectively travel to a conclusion, is not asserted, at least divergence is affirmed. "Experience," says Balmez, in his *Fundamental Philosophy*, "has shown our understanding to be guided by no one of the considerations made by philosophers; its assent when it is accompanied by the greatest certainty, is a spontaneous process of natural instinct, not of logical combinations or ratiocinations." The difficulty here raised may be answered by a few explanations as to facts.

(*a*) When the theoretical account of a case is obviously such as does not take in all the circumstances, then, in practice, we do not follow out the mere dictate of theory. A mathematical formula tells how to point a cannon so as exactly to hit a mark, on the supposition that there is no atmospheric resistance, and no deflecting power in a whirlwind that is blowing. What divergence is there between theory and practice, if the gunner calculates by rule of thumb the disturbing elements, which are too unsettled to allow of theoretic determination? Again, a physician has a scientific theory about the effect of a certain drug on a limited set of conditions within the human body: but aware that these conditions are complicated by many others, which he cannot distinctly formulate, he makes a rough allowance for these last on empirical grounds. Often scientific results are known to be only approximative; and scientific men know how to relax the rigour of these terms to meet refractory cases. One reply to Mr. Stallo's attack on scientific theory was made precisely on this ground, that physicists use "attraction," "fluid," "atom," "potential energy," with a recognised elasticity of meaning, for which only the experienced worker in science can make due allowances. This is an acknowledgment that science is imperfect, but no acknowledgment that it is not in accord with practice: it goes along with practice as far as the length of its own tether will permit. So, too, when it is said that philosophy travels one road, common sense another, it should rather be said, that philosophy is not co-extensive with all practical discoveries, in many of which we know *that* things are, without knowing *how* they are.

(*b*) We should be quite unable to get on in life, if on every occasion, when we wanted a conclusion, we had to go through, in order, all the steps which logically lead up to that conclusion. By dint of habit our mental associations become very nimble, and partly as a matter of direct memory, partly by the aid of dimly suggested inferences, our course is expedited. Whereas the full number of steps are A, B, C, D, E, we seem to go at once from A to E. Some affirm that we do actually pass through B, C, D, but so rapidly as not to advert to the fact; others say that A may have become immediately associated in memory with E, though originally the intermediate stages had to be traversed. At any rate, the impression left is, that the mind takes short cuts to its ends, and that occasionally our conclusions come first, and our premisses, if they come at all, follow afterwards. Instead of being in the case of Dogberry, when he said, " 'Tis already proved you are guilty, and it will go nigh to being thought so soon," we are in a position of saying, " the conclusion is already drawn, and it will go nigh to being proved soon." Something like the strange process which Alice heard recommended in Wonderland, seems to belong likewise to Plain-man's-land, "sentence first, and verdict afterwards."

The account of the process has already been briefly given, but may be repeated with a slight change of words. The mind has gone through much experience, and much labour, in arranging its contents. Many immediate judgments, many syllogisms have been made. As a consequence there is left an orderly register of results; and often a thought gives or seems to give, by direct suggestion, what was originally connected with it through many intervening links. Whether these links are momentarily revived in the memory, but so momentarily as to escape the detection of conscious analysis, need not here concern us; it is enough if we can give an acceptable account of the apparently irrational, or non-rational process whereby reason seems to outrun itself, and to decide before it has the motives. We may add, in this connexion, the theory of Dr. Maudsley, where he explains some of those cases, in which what we are convinced are new matters of thought, nevertheless put on the air of old recollections. He supposes the mind to reach a result before the conscious attention is directed to the process; so that, when consciousness is fully roused, the object seems familiar. In this way the conclusion would appear to anticipate the premisses. The *quasi*-automatic process, however, is always amenable to the judgment of deliberate reflexion, by which it has often to be corrected. A ludicrous instance of inference by rapid association is given in Herodotus, in his story of the revolted slaves, who after repulsing armed attacks, fled when their masters issued out against them with that familiar weapon, the whip. Logical reflexion, if the poor wretches had been capable of it, would have been useful. Thus logic retains her position as the friend and helper of spontaneous reasoning; a position which is accorded to her even by Messrs. Mill, Lewes, and Spencer, who fully admit the use of the syllogism as a "verifying process."

The doctrine above laid down enables us to meet what to the unprepared might seem difficulties, of which a specimen or two shall now be added. "While we assume," says Mr. Sully,[1] "that in

---

[1] *Outlines of Psychology*, c. iii. Reasoning.

reasoning the mind passes from premisses to conclusion, we must remember that this does not answer the actual order of mental events in many, and perhaps in the majority of instances. The conclusion presents itself first, and the ground, premiss, or reason, when it distinctly arises in the mind at all, recurs rather as an after-thought, and by the suggestive force of the similarity between the new case and the old." Mr. Spencer[2] has remarks to the same effect. He says that we go straight from a perceived stone to its lines of cleavage, and do not travel round by the syllogism, "all crystals have lines of cleavage; this stone is a crystal; therefore it has lines of cleavage."

So far from resenting such objections, we welcome them, as helping us to clear up our own conceptions, and as calling our attention to the very important fact, that our mental store does not consist of ideas, isolated like atoms, or standing in rows like words in a dictionary. Rather our ideas make up a sort of organically united whole, one idea developing by epigenesis upon another, somewhat after the analogy of cells in a plant or animal. The analogy is only an analogy, but it is a help for our understanding to conceive, under these figures, processes, the precise nature of which will always be for us a mystery. Goethe compares the union of our mental conceptions to a subtle weaving of many threads together into patterns which gradually display themselves:

> The web of thought, we may assume,
> Is like some triumph of the loom,

[2] *Psychology*, Part II. c. viii. § 305.

>   Where one small simple treadle starts
>   A thousand threads to motion,—where
>   A flying shuttle shoots and darts,
>   Now over here, now under there.
>   We look, but see not how, so fast
>   Thread blends with thread, and twines, and mixes,
>   When lo! one single stroke at last
>   The thousand combinations fixes.³

(c) As too much attention concentrated on the bodily functions may derange them, and as even the simple process of jumping a ditch may fail from excess of care to do it neatly; so an attempt to think out a question in strict philosophic form may deaden or misguide the energies of thought. But these facts argue no essential want of convergence between the spontaneous and the systematized process; the two may be mutually helpful, and each has besides its own peculiar place. Let them combine where they usefully can, and keep apart where combination is detrimental. This is the substantial settlement of the matter; and it meets any such case as that of Sir Walter Scott, who found it sometimes an aid to his progress in a novel, if he began to read a book on some other subject. The desired train of thought, as if jealous of a rival, came in to dispossess the ideas given by the book; just as in a parallel case church-goers involuntarily recall, within the sacred walls, the fact which they tried in vain to recover outside.

3. To put in the limits within which a doctrine

---

³ *Faust*, translated by Theodore Martin, Act II. Scene 1, p. 89. See too Mansel's criticisms upon Locke's "*simple* ideas." (Prolegom. *Log.* c. vi. p. 185.)

is meant to be accepted, often saves a world of misconstruction ; and the present instance is one calling for a statement of limitations.

First, no account is taken of grace and of supernatural revelation, though both are facts. What we call revelation is of rarer occurrence, and vouchsafed only to the favoured few: but unless the Church is to give in to Pelagius, and to those who go further than ever Pelagius went in the direction of naturalism, she must maintain that Christians are in constant receipt of illuminations by grace from above, both as to their faith and as to their guidance in conduct.

Besides the supernatural mysticism treated of by the Pseudo-Dionysius, his commentator Maximus, St. Bernard, Hugo and Richard of St. Victor, St. Bonaventure, Gerson, and pious writers who have not been professed theologians, there is asserted also a sort of natural mysticism. This we must make over to the Society for Psychical Research, for it cannot be reduced, by our present knowledge, to logical system : whereas the truths that can be so reduced suffice for a Philosophy of Certitude.

### Addenda.

(1) The Tractarian movement, at Oxford, offers some instructive contrasts between the mind which holds that thought can be rigorously carried on, and the mind that distrusts philosophy. In the notice of the death of the late Dr. Ward, a leader in *The Times* remarked pointedly upon the circumstance, that in his University

days he was a noted stickler for logic ; "whereas," adds the writer, "most people are content to say as much as meets the occasion, in the blandest form and in the pleasantest tone. Logic is not much required for the dinner table or on the platform."

Before bringing forward the contrast between Dr. Ward and other men at Oxford, it is worth while inserting an illustration precisely of this "bland form and pleasant tone" of the illogician. "One peculiar defect of mine," confesses or boasts M. Renan,[1] "has more than once been injurious to my prospects in life. This is my indecision of character, which often leads me into positions, from which I have a great difficulty in extricating myself. This defect is further complicated by a good quality, which often leads me into as many difficulties as the most serious of my defects. I have never been able to do anything which would give pain to any one. . . . In talking and in letter-writing I am at times singularly weak. With the exception of a select few, between whom and myself there is a bond of intellectual brotherhood, I say to people just what I think is likely to please them. With an inveterate habit of being over polite, I am anxious to detect what the person I am talking with would like me to say. My attention, when I am conversing with any one, is engrossed in trying to guess his ideas, and from excess of deference to anticipate him in the expression of them. My correspondence will be a disgrace to me, if it is published after my death." From this charge of extreme complaisance he excepts his published works; but they too must be affected by certain qualities which shall be added for the completion of the picture. "By mere force of things and despite my conscientious efforts

[1] *Recollections of my Youth*, the Part entitled, *St. Renan*, p. 65. (English Translation.)

to the contrary, I am a member of the romantic school, protesting against romanticism; a Utopian inculcating the doctrine of half-measures; an idealist unsuccessfully endeavouring to pass muster for a realist, a tissue of contradictions resembling the double-natured hircocerf of scholasticism. One of my two halves must have been busy demolishing the other half, and it was well said by that keen observer, M. Challemel-Lacour, he feels like a woman and acts like a child. I have no reason to complain of such being the case, as this actual constitution has procured for me one of the keenest intellectual joys a man can taste." That will do for M. Renan; now for Dr. Ward's more immediate contrast.

Again the risk of doing an injustice is avoided by our being able to quote an autobiographical sketch, of which the responsibility lies with the subject. Speaking of his part in the Oxford movement Mr. T. Mozley says:[2] " Why did I go so far in the movement, and why did I go no further? Why enter upon arguments, and not accept their conclusions? Why advance to stand still, and in doing so commit myself to a final retreat? The reasons of this lame and impotent conclusion lay within myself, wide apart from the great controversy in which I was but an intruder. I was never really serious, in a sober, business-like way. *I had neither the power nor the will to enter into any great argument, with the resolution to accept the legitimate conclusion.* Even when I was sacrificing my days, my strength, my means, my prospects, my peace and quiet, all I had, to the cause, it was an earthly contest not a spiritual one. It occupied me, it excited me, it gratified my vanity, it soothed my self-complacency, it identified me in what I honestly believed to be a very grand crusade, it offered me the hope of contributing to very grand achievements. But good as the

[2] *Reminiscences of Oriel*, Vol. II. c. cx. p. 270. Compare c. cxvi.

cause might be, and considerable as my part might be in it, I was never the better man for it."

If it may be permitted to allude to yet a third autobiography, we will mention the *Memoirs of Mark Pattison*, who tells how, having engaged in the Tractarian movement, he ended by diverting his thoughts from it to scientific ideas, and his Tractarianism succumbed, not to argument, but to " inanition "—died of starvation.

In the order of God's providence these things are " written for our instruction," that so far as we have the opportunity and the need, we may train our minds to follow a more rigorous method of thinking. It is suggestive in the course of reading, to notice who are the authors who express their contempt for philosophic system, and who claim a free range for thinking as they fancy. A significant list could be drawn up, in which the much-belauded Goethe would stand as a warning example; though not all would recognise that his want of hold upon systematic truth was a calamity (Goethe, *Sein Leben und Seine Werke*, von Alexander Baumgartner, S.J., Vol. I. pp. 27, 28).

(2) In behalf of the view that human thought is essentially loose and inaccurate, it may be argued that philosophy has shown the same characters in the formation of grammatical forms. Far from having a strict propriety in them, many are traced back to bad analogies, to pieces of clumsiness, and to downright blunders; so that a man who has had a little insight into the origin of some usages, is not much inclined, at this late hour, to do vigorous battle in the cause of a fancied purism against established usage. If it be asked, Why may not thought have its inner anomalies of a like character? the reply is ready at once, Because thought is not language. The latter is made up of conventional signs, which may very well have had an illogical origin;

whereas thought is no conventional sign, but the most natural of all natural signs. Thought, if anomalous, is simply undone.

(3) What is called "unconscious thought," by the aid of which many of the mind's gathered materials are supposed to be automatically arranged, will be considered in the chapter on consciousness. It may very well be that certain cerebral changes go on unconsciously, which yet are most useful or needful for the clearing up and arranging of thoughts; but whatever these processes, the final outcome will have to be judged on conscious principles before it can reasonably be pronounced true or false.

(4) In reference to what has been said about the reasonable defensibility of all vital truths, we may profitably quote a decree of the Congregation of the Index, of June 11, 1855: "Reason can establish with certainty the existence of God, the spiritual nature of the soul, and the freedom of man's will."[3]

[3] "Ratio Dei existentiam, animæ spiritualitatem, hominis libertatem, cum certitudine probare potest."

# CHAPTER VIII.

### UNIVERSAL SCEPTICISM.

*Synopsis.*
1. Division of scepticism. (*a*) Dogmatic scepticism. (*b*) Non-dogmatic scepticism.
2. Other sciences may refute themselves, but not so the philosophy of certitude.
3. Scepticism is incapable of giving the promised rest from anxious questionings.
4. A word on Hume, the father of English scepticism.

*Addenda.*

THE next subject may be introduced by a character described in the *Essays of Elia:* " He hath been heard to deny that there exists such a faculty at all in man as reason, and wondereth how men first came to have the conceit of it—enforcing his negation with all the might of reasoning he is master of. He has some speculative notions against laughter, and will maintain that laughter is not natural to him—when peradventure the next moment his lungs will crow like chanticleer. It was he who said, upon seeing the Eton boys at play in their grounds, What a pity to think that these fine, ingenuous lads, in a few years, will all be changed into frivolous Members of Parliament!"

The character of the sceptic has always been one of which jokers have made capital, and Lamb has taken his turn in the mockery. Against the possible existence of a complete sceptic, as a fact of real life, those who themselves have been supposed to be far gone in the same malady, have clearly pronounced. Hume[1] says that such a being is imaginary, for speculative doubts give way utterly before the pressure of practical life. Rather than have sceptics argued with, he would have them left alone, lest opposition should feed that perversity, which, abandoned to itself, would perish of its own weakness.

1. Nevertheless we must do a little in the way of argument, if not with sceptics, then against scepticism; and we may take, as a division of the matter, what is given by Sextus Empiricus. His account may not be historically accurate, but at least it furnishes two convenient headings under which to confute scepticism. "Many persons," writes Sextus,[2] "confound the philosophy of the Academy with that of the Sceptics. But although the disciples of the New Academy declare that all things are incomprehensible, yet they are distinguished from the Pyrrhonists in this very dogmatism. The Academicians affirm that all things are incomprehensible—the Sceptics do not affirm even that. Moreover the Sceptics consider all perceptions perfectly equal as to the faithfulness of their testimony: the Academicians distinguish between probable and improbable perception." Here we have the suggestion of the par-

---

[1] *Inquiry*, Part II. sec. xii. in fine, et alibi passim.
[2] Ueberweg's *History of Philosophy*, Vol. I. Second Period of Greek Philosophy, § 60, p. 213. (English Translation.)

tition of sceptics into dogmatic and non-dogmatic; those who make a dogma of their very doubt, saying that the one certainty is the uncertainty of all human opinions, and those who abstain from claiming even this one certitude. It should be observed, however, that unless a sceptic were extra strange among a class of strange beings, he would hardly pretend to doubt the facts of his own consciousness —that he had those feelings which he experienced. What he would question would be the objective reality of his thoughts, not his subjective states as such.

(a) The fatal act of the dogmatic sceptics is their profession to have strictly proved their conclusion, and to hold it positively as a valid inference. Being, as John of Salisbury describes them, " Men whose whole endeavour is to prove that they know nothing,"[3] they elaborately argue out their case, and make quite a system of their views.

Now their conclusion is either proved or not. If it is not proved, then they have failed in their main object: if it is proved, then the many facts and principles, which went to build up the proof, are thereby declared invalid; for they imply a large mass of human certitudes. In the premisses the sceptics appeal to observed facts, within and without their own persons: these facts they discuss in connexion with the principles of reason, and draw inferences. Do they accept the observations and the principles as valid? If so, theirs cannot be the final conclusion to gather from them, for this con-

[3] "Quorum labor in eo versatur, ne quid sciant."

clusion, when drawn, at once turns round on the premisses and says, "Out upon you, you vile incapables, you are yourselves suspects, and can lead only to suspicious conclusions." The premisses retort, "That reproach does not come well from you." To affirm positively the invalidity of all reasoning, supposes a mind capable of a number of valid decisions: the one dogma of scepticism can never stand alone.

The mistake of the dogmatic sceptics seems to be some lurking notion, that argument ending in denial need not imply fixed principles, but may be like simple nescience. Possibly they look to some false analogy, like that of a drunken man, with just sense enough left to see that he cannot transact business, and had better seek retirement; or, again, like that of an insane man, who sufficiently perceives his own state, to beg that he may be taken to an asylum; or, lastly, like that of a constitutionally feeble intelligence aware of its own imbecillity. In the inebriate, in the insane, in the imbecile, there may be intermittent gleams of right reason, and the examples form no true parallel to the case, in support of which they are supposed to be adduced. A light shining faintly and fitfully through a cloud, does not illustrate the paradox of a light showing itself to be absolute darkness.

The position of the dogmatic sceptics, when they have done and said all, remains worse than that of the dumb man who tries to speak out and declare his own condition: or that of those who had to solve the old puzzle, how to believe, on a man's own testimony, that he is an unmitigated liar. Concerning this latter knotty point, we are told that Chrysippus wrote six volumes, and that Philetas so overtaxed his energies as to die of consumption and deserve the epitaph :

Stranger, Philetas am I ; that fallacy called " The Deceiver," Killed me, and here I sleep, wearied of lying awake.[4]

The problem of dogmatic scepticism is calculated to prove equally killing.

The dogmatic sceptic need not maintain his power to determine grades of probability ; but since the New Academicians are said to have added this burden to their charge, and since the matter, when investigated, throws more light upon the position of scepticism, we shall do well to put in a word about the sceptic's probabilities. When a probability is declared by moralists to justify a certain course of conduct, they still admit that an action, only probably permissible, would be illicit : for a man is not allowed to act at a venture. But falling back upon a principle which they regard not as merely probable, but as certain, namely, that under some circumstances, where the obligation is not clear, it is no obligation at all, they succeed in establishing the maxim, *Qui probabiliter agit, tuto agit.* The safety is not simply in the probability, but in the certainty as to how they may act, where what stands in the way of action is only a probability against its being

---

[4] Ξεῖνε, Φιλητάς εἰμι· λόγων ὁ ψευδόμενός με
"Ὤλεσε, καὶ νυκτῶν φροντίδες ἑσπέριοι.

# UNIVERSAL SCEPTICISM.                139

allowed.⁵ What is thus illustrated in morals has an analogous illustration in intellectual matters. Here also a probability requires the aid of some certainty. To calculate probabilities and assign their several grades, needs a mind which knows, by its experience, how to discriminate the state of doubt from the state of certainty, and which has many certainties whereby to fix the probabilities. It is simply ridiculous for dogmatic sceptics to claim that skill which the Academicians claimed, in the nice adjustment of a scale of probabilities.⁶

(*b*) The non-dogmatic sceptics have the greatest difficulty in describing themselves, for they are not allowed definitively to declare anything, not even their universal scepticism. One Greek philosopher tried to evade the difficulty by pointing out his meaning with his finger; but there is a limit to communication by this means, nor does the device exactly fulfil its purpose. The boasted "dumbness" or "suspension of judgment"⁷ cannot be maintained. Indeed, the non-dogmatic sceptics make long discourses and write big books, in spite of the obvious objection, that in their case there is special force in the malicious wish, "O that mine enemy had

---

⁵ Mill is a probabilist in his *Subjection of Women*, p. 3. "The *a priori* presumption is in favour of freedom. Those who deny in women any privilege rightly allowed to men, must be held to the strictest proof of their case, so as to exclude all doubt."

⁶ Hume teaches "that all our knowledge resolves itself into probability:" and that he "had almost said this was certain," but refrains on reflexion "that it must reduce itself, as well as every other reasoning, and from knowledge degenerate into probability." (*Treatise*, Bk. I. Part IV. sec. 1.)

⁷ ἀφασία or ἐποχή.

written a book." To their books they try to sign the name of their school of thought. Now without any insult to them, let us, merely as an illustration, compare their procedure with the case of the animal that is really an ass; how is the poor brute to write itself down accordingly? A bray is about the best sign it can give as "its mark." Similarly, a non-dogmatic sceptic, who for reasons set down in his book, takes up his position, is forbidden, by the very terms of his profession, to say positively what his intellectual stand-point is. To say "I am a non-dogmatic sceptic," would be as clear a piece of dogmatism as to say, "I am a dogmatic sceptic;" for it would imply that dogmatic scepticism was wrong, and that the right attitude was to be without any affirmation whatever. Yet so to teach is itself an affirmation, resting on many others.

Briefly, the non-dogmatic sceptic either keeps to his profession of inability to speak (ἀφασία) and affirms nothing, in which case there is nothing to refute, but at most we can complain of faculties unused; or else, breaking loose from his engagements, he makes an affirmation, and so refutes himself. This suffices to end the general attack on the position of universal scepticism: attacks in detail must follow afterwards, as occasions successively offer themselves.

2. The peculiar position of the *Philosophy of Certitude* is not appreciated by the sceptic. Another science might be held to furnish its own refutation by presenting manifest contradictions; but there cannot, in the same way, be a sceptical refutation of

the *Philosophy of Certitude* by that philosophy itself, for there would no longer be an umpire left to give the award of victory or defeat. If in a theory of light the application to phenomena of reflexion and refraction belies the application to phenomena of diffraction, then a mind is still by to judge of the contradiction, and of its fatal consequences to the theory: but if the very mind itself is to be proved essentially contradictory, how is it to establish the result? Mill[8] seems to share with the sceptics their want of appreciation for the position, when he writes: " If the reality of thought can be subverted, is there any particular enormity in doing it by the means of thought itself? In what other way can we imagine it to be done?" Surely this argument is fallacious: because there is repugnancy in supposing anything but thought to work a certain effect, therefore there can be no repugnancy in supposing thought to work it. Mill, however, continues unembarrassed: " If it be true that thought is an invalid process, what better *proof* can be given, than that we could in thinking arrive at the conclusion, that our thoughts are not to be trusted? The scepticism would be complete even as to the validity of its own want of belief." As men, after execution, cannot sign a document testifying that sentence has been carried out, neither can reason sign a valid testification to her own proof of her own universal invalidity. A man may with one eye see that the other is hopelessly injured, whether he use a mirror for

[8] *Examination of Sir W. Hamilton's Philosophy*, c. ix. pp. 132, seq. (2nd Ed.)

the purpose, or employ the faculty which a celebrated Greek philosopher is said to have possessed, of making the eyes converge till they looked into one another; but a single blind eye will never literally see its own destruction. Mill, though sometimes patronizing the man who never believed in dreams because he dreamt that he must not, yet in a better frame of mind himself confesses, that "denying all knowledge is denying none."

Hamilton is another who has let himself be caught in the same trap, when he puts a hypothesis which he ought to have seen to be contradictory: "The mendacity of consciousness is *proved* if its data, immediately in themselves or mediately in their consequences, be shown to stand in mutual contradiction." Glad to agree with one from whom we often differ, we may let Mr. Spencer[9] answer here: "It is useless to say that consciousness is to be presumed trustworthy until proved mendacious. It cannot be proved mendacious in this primordial act. Nay, more, the very thing supposed to be proved cannot be expressed without recognizing the primordial act as valid; since, unless we accept the verdict of consciousness that they differ, mendacity and trustworthiness become identical," or at least not distinguishable. "Process and product of reasoning both disappear in the absence of this assumption."

3. Scepticism, being so clearly a sin against the right use of intelligence, could not lawfully be paid as the price of rest from all anxious questionings,

---

[9] *First Principles*, Part II. c. ii. § 41.

even if the bargain were possible. But it is not possible. For the complete sceptic is, as Mill[10] says, "an imaginary being," never to be actualized: while such scepticism as man can actualize, certainly does not bring the promised quietude, or "absence of disturbance" (ἀταραξία). The case is as with drink. If drink could perfectly drown care, still we ought not to turn drunkards: besides, drink does not effectually drown care, for it brings in its train alternations of great suffering. Our true peace is to be sought in a right use of that reason, in which is the great root of our responsibility, and the alternative source of our highest happiness or misery. And when we remember that our reason is not our own independent property, but a gift—an entrusted talent —we shall be far indeed from calling her calumniously, with Bayle, "the old destroyer," "the cloud-gatherer," and far from adopting the pernicious sentiment of the verses :

> Thinking is but an idle waste of thought,
> And nought is everything, and everything is nought.

Rather we shall recoil from intellectual nihilism as a Russian Czar abhors social nihilism : for the loss of all belief in intellect tends to paralyze action, and to take the energy out of life by robbing it of its hope.

4. Unfortunately, though not going under the name of sceptics, but rather of agnostics, there is a large party of our philosophers in this country, who are pledged to the fundamental principles of

[10] *Examination*, c. ix. in initio.

scepticism in accepting substantially the doctrine of Hume. The irresoluteness of their chief might warn them to distrust him. While his principles are sceptical, he claims, in spite of them, to retain his belief: he finds comfort in setting up practice against theory, and declares, "as an agent I am not a sceptic:" he adds that there is no real sceptic. Ferrier goes so far as to suppose that Hume was not serious in his work, but was aiming at the *reductio ad absurdum* of the philosophic principles prevalent in the England of his day. Dr. Symon, taking up a like view, says that Hume was "merely and undisguisedly sarcastic, and in jest, never in earnest, when he wrote on metaphysics." Even one who has no little sympathy with Hume, Mr. Bain,[11] declares, "As he was a man fond of literary effects, as well as of speculation, we do not always know when he is in earnest." The fair estimate of Hume seems to be, that he is not quite as bad as he appears: that many of his efforts were tentative: that he began to destroy, and then, alarmed at his own vandalism, set himself to build up again: that his avowed principles were sceptical, but that he dared not, and could not, push them to their extreme conclusions. Hamilton tries, but not apparently with full success, to save Hume's consistency by the plea that to arrive at an inconsistency was the very object of his aim, it being "the triumph of scepticism to show that speculation and practice are irreconcilable."[12] In agreement with

[11] *Mental Science*, Bk. II. c. vii.
[12] *Hamilton's Reid*, p. 437. Cf. pp. 129, 144, 489.

this view stands Hume's oft-quoted account of Berkeley's sceptical arguments, that they " admit of no answer, and produce no conviction."[13] Finally, Hume's recent editor, Professor Green, decides that " when we get behind the mask of concession to popular prejudice, partly ironical, partly due to his undoubted vanity, we find much more of the ancient sceptic than of the positive philosopher."[14] At any rate this is certain, that Hume should have no influence with a well balanced mind, which reverences itself as the greatest natural power upon earth, and as the only means of entering into moral communication with the highest Power of all. Mind is our mightiest possession : νοῦς πάντα κρατεῖ.

### Addenda.

A posthumous work, sent out in the name of the famous French Bishop, Huet, is a combination of the tenets of non-dogmatic scepticism with the assertion of the dogmatically sceptical academics, that there are degrees of probability in our opinions about things. There were not wanting in France, about his time, abundant seeds of scepticism, diffused by Montaigne, Charron, Francis Sanchez, Bayle, Pascal, and others. Furthermore Huet might feel that he was not the first prelate to put forth the style of doctrine which he was maintaining ; for about two centuries before, Cardinal Nicholas of Cusa, had written his works, *De Docta Ignorantia*, and *De Conjecturis*, to show the impotence of human reason, and to affirm the need of some sort of

---

[13] *Inquiry*, Part I. sec. xii.
[14] Introduction, § 202. See Hume's account of his own feelings, *Treatise*, Bk. I. Part IV. sec. vii.

K

intuition of God. Huet's *Feebleness of the Human Mind* appeals to isolated passages of Scripture, and of the Fathers, which seem, in their naked form, to give some countenance to the view, that man's intellect is incompetent, and that knowledge must be given from on high. But these utterances, separated from their original accompaniments, ought to have been taken in their context, and with the light shed upon them from other passages, expressly declaring the prerogatives of human reason. As to Scripture, it is its style not to put in qualifying clauses, but to take one side of truth and speak for the time as though this were the only side. Now faith alone, now works alone, are spoken of as efficacious : the full truth being, when its elements are fused together, that works done in faith are requisite. The Fathers likewise do not think it always needful cautiously to balance one truth by its counterpart.

Huet thus endeavours to state his position of non-dogmatic scepticism : "In saying that nought is either true or false, I enunciate a proposition which refutes itself, as it is not excepted from the general law, which says that nothing is either true or false."[1] About sceptical arguments in proof of the position, he says : "They subvert other propositions, while subverting themselves, it is for this sole purpose they are enunciated, and not with a view to proving them."[2] Other authors make the same statement in another shape, saying that scepticism is like a drug which purges out everything, itself included.

[1] "Lorsque je dis qu'il n'y a rien de vrai ni de faux, cette proposition s'enferme elle même, et elle n'est pas exceptée de la loi générale qui prononce qu'il n'y a rien de vrai ni de faux." (*De La Faiblesse de L'Ésprit Humain*, Liv. III. ch. xiii.)

[2] "Elles détruisent les autres propositions, en se détruisant elles-mêmes ; car c'est seulement pour cela qu'on les emploie et non pour les établir." (*Ib.*)

Huet places what he conceives to be the superiority of his stand-point over that of ordinary mortals in this: "They know nothing, and we know nothing, though we feel uncertain about our nescience. Further, while they do not question our probability, we do deny to them the possession of the truth which they seek after."[3] The case is not so at all: for Huet cannot more vigorously deny to us our certitudes, than we deny to him his probabilities, if the probabilities are to be calculated on his principles.

[3] "Ils ne savent rien et nous le savons, quoique incertainement et en doutant. De plus, il ne nous contestent la vraisemblance que nous suivons, et nous leur refusons la vérité qu'ils recherchent."

# CHAPTER IX.

### CARTESIAN DOUBT.

*Synopsis.*
  1. The methodic doubt of Descartes as distinguished from mere scepticism.
  2. The plausible part of Descartes.
  3. Passages in his works whence to gather the substance of his method.
  4. The destructive part of his work.
  5. It falls into the principles of universal scepticism, and makes the future work of construction logically impossible.
  6. The constructive part itself.
  7. General estimate of Descartes.

*Addenda.*

1. THE doubters with whom we have just been dealing make doubt their final goal, they doubt and rest there: but we have now to deal with a universal doubt which is supposed to be a means of helping on the mind towards well-assured knowledge. Hence it is called *methodic* doubt, as being only a way, or rather part of a way, to an end, not an end in itself. Descartes who, it should be remembered, gives warning that his system is dangerous for all but the few, is the deviser of this method of doubt, which has won for him more credit with some people than close investigation of its merits will bear out. The fact is, Descartes says many things

that are either quite true, or contain an obvious element of truth; and, in his replies to objections, he may seem to get over certain difficulties which, if reference were made back to his system, would be found to be insuperable. But of course few readers go to the trouble of making such reference, and so the author is the gainer. Even a well-informed writer like Hamilton,[1] speaks of the error of Descartes as accidental rather than substantial; whereas his error is substantial and the admixture of truth accidental. There are other critics who, to less attentive readers, may appear to approve, in the main, of Descartes, yet who, if read more carefully, will be found to disagree with him fundamentally. Instances are Balmez, Sanseverino, and Tongiorgi. However, our business is much more to refute the popular version of Cartesianism, than to score a victory over one long since dead and beyond the reach of our weapons; so that to us it is a matter of small consequence, whether a wide collation of passages might not do something to mitigate the crudeness of the system, when taken in outline.

2. What the snatch-and-away class of readers would seize upon in Descartes is just what is most plausible and insidious. The surface of his doctrine looks fair, and the prominent parts are easily grasped. He finds that his mind is like a basket containing apples, good and bad: and he proposes to empty the whole out, and put back only the good. Certainly a very natural thing to do, if the mind is a basket of apples. But so patently is the mind not

[1] *Logic*, Vol. IV. Lecture xxix. p. 91.

a basket of apples, that a directly opposite course of action has suggested itself to others. Thus Cardinal Newman has declarations to the effect that, if he were driven to choose between the two extreme alternatives, he would rather begin by holding all present beliefs, and gradually letting go the untenable, than start with the clean sweep made by universal doubt. And this process Wundt actually recommends, so far as he teaches, that instead of beginning from the idealist point of view, men should first hold their ideas to be real: then they should eliminate what can be shown to be merely subjective, and keep the residue as objective. Thus the analogy of the basket is catching indeed to an average reader; but catching in the way of that now forbidden article, the man-trap.

3. In three different places of his works, Descartes describes the successive steps of his system; yet to inquire what precisely this system is, seems hardly to enter seriously into the minds of ordinary retailers of philosophic opinions. Perhaps they are secretly led by the principle which we have seen Mr. Pollock avow, namely, that "systems" are nothing, but a few "vital ideas" everything. With regard to Descartes, any one who will carefully compare his threefold account of his system, will be quite convinced that the author had not steadily made up his mind how the several steps in the progress were to succeed each other. The *Discourse on Method*, Part IV., the *Meditations*, especially the first, and the *Principia* (Part I. in initio), would not quietly fuse together into a *Summa*, though they are

meant to be three descriptions of one leading process. However, in the destructive part of this process, Descartes is pretty uniform : and it is this part chiefly which we must assail, destroying the destroyer.

4. The philosopher soliloquises somewhat in this strain : I, being now in the maturity of my faculties, find that the formation of my opinions has hitherto been not at all critically conducted ; and whereas it would be endless to test each of my beliefs separately, therefore I must aim at some comprehensive method. Recurring to my reasons for dissatisfaction, I find that my senses have often deceived me, and therefore as means of knowledge they are to be suspected : which suspicion is immediately extended to the rest of my knowledge, so far as it has its beginning in the senses. But my intellect itself is open to direct assault : it too has been deceived in matters when I felt quite sure, and I can doubt even about mathematical truths, which are considered as types of clearness. Next as to grounds of misgiving which are extrinsic to my own faculties, sensitive and intellectual; whence have I these faculties? I am told that I have them from an Omnipotent Creator: and if He is Omnipotent, He can do all things, and consequently He can make me essentially a creature of delusions. Or suppose I am the work of a maker less than omnipotent; then all the more likely is the less powerful maker to have made me ill. But perhaps this is irreverent : so let us suppose it is some evil spirit that is perpetually turning me to mockery. Thus on all sides I find my very faculties untrust-

worthy, and trying to doubt, I can doubt the existence of my body and its senses, of earth and heaven: "and finally I am driven to admit that there is nothing of what I previously believed which I cannot in some way doubt: and this not lightly and inconsiderately, but because of very strong and well-weighed reasons."

It is not extravagant to hope the reader will allow, that the way to criticise the above "method" is not simply to look out for some stray "vital ideas" which it may contain, but to look to the whole method of which a part has just been sketched, and ask, can the proposed whole admit of that part. Descartes is not arguing in behalf of permanent doubt: else he would be one of the dogmatic sceptics refuted in the last chapter: he is arguing for doubt as a preliminary to certitude, and this fact is vital to his *system*, whatever may be the "vitality of ideas" out of systematic connexion with each other. Now as a system Descartes' method fails, if his principles of destruction are inconsistent with any subsequently applied principles of reconstruction. He first doubts in order afterwards to be certain: he does not indeed try to draw certitude out of doubt itself; but he does try to start from a state of doubt on the way to certitude. Hume[a] and Reid agree that he has so buried himself beneath the ruins of the edifice he has pulled down, that rebuilding is beyond the power of the utterly crushed enterpriser.

5. If it were necessary, for purposes of refuting

[a] *Inquiry*, Part I. sec. xii. in initio.

his "method," to follow Descartes into all the details of his arguments, we should require at once to enter upon such special subjects as the trustworthiness of the senses, the nature of mathematical truths, the nature of necessary truth, the regulation of Divine omnipotence by Divine wisdom and goodness, the permission of evil, the powers of wicked spirits in face of a Provident Ruler, and other large questions. But there is a shorter way: Descartes falls into the inconsistencies of the universal sceptics, and is logically forced to abide with those in whose company he is unwilling to remain. He professes to be able, "seriously and for well-weighed reasons," to doubt the validity of his faculties, and truths which present themselves to his mind with the force of evidence. Out of such doubt there is no rescue. A man so circumstanced has no right even to his " I think, therefore I exist " (*Cogito, ergo sum*); and if he says that on this point doubt is impossible, he says so only by revoking what he had said before; for if his whole nature may be radically delusive, it may be delusive here. He says the doubter cannot doubt his own existence: but neither can the doubter doubt consistently the validity of his own faculties and of evident propositions. Some have so bemuddled themselves that they have felt alarmed as to their own existence; and a large system of pantheism denies the reality of the separate *Ego*. If this bemuddlement is a degree worse than that of Descartes, the question is only one of degree, not of kind. It is substantially the same kind of evidence which testifies that I exist, and that what

I know, I know, or that my faculties are veracious. A man may and must start from ignorance, and by the experience of his intellectual life first discover, empirically, that he is an intelligent being: also a man may gradually test by experience that he is waking up from a dream or from a delirium. But no man, from the position of what Descartes styles the proved suspiciousness of his very power of knowing anything, can coolly go on to use his suspected faculties as witnesses in their own behalf, when they say *Cogito, ergo sum*. The only irrefragability of Descartes, at this point, is the convincing evidence of his maxim on other principles than the Cartesian, not on Cartesian principles.

There is, however, one point stated in the last paragraph which ought not to be left without further notice; and it is that some defence may apparently be made for Descartes, inasmuch as he places the certainty of self above the certainty of ordinary truths which are immediately evident. A large number of philosophers have remarked that our own states of consciousness, and a knowledge of *some kind* of self, are matters beyond all question: whereas at least a question may be raised as to whether our thoughts in general stand for any objects beyond themselves. The absolute unquestionableness on the one side, and the possible questionableness on the other, seem at first sight to rest on a well-grounded distinction: but closer inspection will not bear out first impressions. For if we push scepticism concerning truths other than the truth of our consciously modified self to their logical conclusions, we shall find ourselves reduced to the inability of making any certain declaration whatever. We shall be as ill off as Mill[3] when he admitted the *necessity of deductions* from axiomatic truths, but denied the *necessity of the axioms*: as though the evidence for one were not as compelling as the evidence for the other, and as though reasoning could have a prerogative over immediate intuition. If Hume[4] is any support we have him as an ally in the present instance; for he denies to Descartes that there is "any original principle which has a prerogative over others," such as the *Cogito, ergo sum* is asserted to have. Allow Descartes' principles to the full, and instead of your fixed certainty that you, the doubter, exist, you will find yourself muttering some verses of Byron, which one sees occasionally quoted:

> So little do we know what we're about in
> This world, that I doubt if doubt itself be doubting.
> O doubt, if thou be'st doubt, for which some take thee,
> But which I doubt extremely, &c.

These expressions are wild utterances, but Descartes has no right to complain of them, and he ought to have realized the startling fact.

Yet so easily do certain minds isolate "vital ideas," that some speak as though the whole onslaught against Descartes, was because he stood up vigorously for the fact of self-existence, as revealed

---

[3] *Logic*, Bk. II. c. vi. § 1.
[4] *Inquiry*, Part I. sec. xii. in initio.

in thought and as a primary cognition! We all stand up in defence of that piece of knowledge; our quarrel is with the previous scepticism. We would wipe out from Descartes' system other things besides, but first of all, that which most strongly characterizes it, its initial stage of universal doubt. Here again the "good easy reader" of reports at second hand, seems to be under the delusion, that Descartes merely said we had reasons for dissatisfaction with our early way of laying up mental stock, and that the stock in hand should, in mature years, be thoroughly overhauled. Descartes teaches a great deal more than that: he claims to have proved, by reasons, that mathematical evidence may be fallacious, and that so may be our very inmost nature. Do not overlook this essential part of the system, if you would be anything like a competent critic: and do not fail to notice how such a beginning is absolutely fatal to further progress.

6. On the principles involved in his "methodic doubt" alone, Descartes would find defence impossible; but he labours under the further disadvantage, that there are principles, in other parts of his philosophy, which serve to cripple him very much, and render it still more difficult for him ever to recover his certitudes. Truth, according to Descartes,[5] rests ultimately on the Divine free-will: and had God so chosen, our necessary truths might have been the reverse of what they are. This is a very different thing from saying, that God could have given to us, or to other beings having our place,

---

[5] *Meditations, Réponses aux Sixièmes Objections*, n. 8.

a palate which enjoyed oil of vitriol, and a stomach which could digest aconite; in all which assertions there is no clear contradiction. But to assert that God could have reversed, not merely physical arrangements, but also metaphysical principles, is to strike at the root of all truth and of all knowledge, and to annihilate the difference between truth and falsehood. Truth is no longer a sacred thing; and that God should use His omnipotence to deceive us, no longer admits of disproof. In fact nothing admits of proof or disproof, for that which both aim at ceases to have a meaning.

As to the constructive part of the Cartesian system, we need only note its futility. In some accounts, next to his first great fact, *Cogito ergo sum*, he places a criterion of truth derived from the experience of this fundamental certitude. This last is accepted because it is contained in clear and distinct ideas: hence is derived the criterion: " That is true which is contained in clear and distinct ideas." But in other places Descartes pronounces the criterion, so obtained, to be invalid—an invalidity which some might suppose him to limit to the external world—until we have settled, that the faculty which has the clear and distinct ideas is from God, who cannot create lying powers of mind. Onward, therefore, to the proof of God's existence Descartes hastens: and argues in a circle, that God exists because our clear ideas affirm it, and our clear ideas validly make the affirmation because God is their voucher. Few who praise Descartes as the philospher of "clear thought," care to look into his theory of "clear ideas:" and from that theory their own opinions are utterly dissentient. Yet it is a fact that often a doctrine cannot be understood till its meaning is made to square with its context, and it is ridiculous to pretend to be in admiring agreement with an author, when really you and he are radically at disagreement, and when he does not decisively know his own mind. As a system Cartesianism is quite without supporters: and this is a fact—a most important fact—which a careful examination cannot fail to reveal to fancied adherents.

The general estimate of Descartes is by some put very high, by others much lower. Buckle, not a great authority on abstract sciences, is quite in the characteristic vein of the *History of Civilization*, when he calls Descartes "the Luther of Philosophy," who "believed, not only that the mind by its own effort could root out its most ancient opinions, but that it could, without fresh aid, build up a new and solid system. It is this extraordinary confidence in the power of the human intellect which gives this philosophy that sublimity which distinguishes it from all other systems." If Buckle had known more of what he was talking about, he would have been checked by the reflexion, that Descartes, in places where he brings forward his half-hearted theory of innate ideas, goes very near, at times, to denying the intellect's power of forming its own conceptions, and to declaring it wholly dependent upon infused ideas; that he takes away from us any natural means of passing from sensations to thoughts; that he makes all our certitude rest on the knowledge of God as the Author of our faculties, whilst this idea of God he makes necessarily dependent on a Divine communication.

The real position of Descartes seems to be, that he brought into prominence some useful doubts and some useful conceptions, which others carried to better issue than he did, and in this respect he not a little resembles Bacon; also that he started some dangerous ideas, which again others carried to worse issue than he did. It is of the latter that Bossuet, himself a sort of Cartesian, wrote: "To conceal nothing from you, I see that a tremendous conflict threatens the Church, under the name of Cartesian philosophy. I see that more than one heresy will spring from its principles, though, as I believe, from their wrong interpretation."[6]

The mathematical services of Descartes were admittedly great, especially his share in the invention of analytical geometry; and in the physical sciences he is quite welcome to whatever honours his friends can vindicate for him; it is only his "methodic doubt" that is here expressly condemned. Yet in regard to science as distinguished from philosophy, it may be noted that Whewell, in his *History of the Inductive Sciences*, lodges against him such charges as, that he misstated the third law of motion; that he claimed to himself discoveries of Galileo and others, which cannot be allowed to one who "did not understand, or would

[6] "Pour ne vous rien dissimuler, je vois un grand combat se preparèr contre l'église, sous le nom de philosophie Cartesienne; je vois naître de son sein, à *mon avis mal entendu*, plus d' une hérésie."

not apply, the laws of motion which he had before him;" that "if we compare Descartes with Galileo, then of the mechanical truths which were easily obtainable in the beginning of the seventeenth century, Galileo took hold of as many, and Descartes of as few, as was possible for a man of genius;" that "in his physical speculations Descartes was often very presumptuous, though not more than half right," that he would not question nature, being ambitious of showing not simply what is, but what must be. These accusations may, or may not be justified, as far as we are concerned; our one great accusation is, that Descartes attempted the impossible, in trying to build up a system after giving positive reasons for the conclusion, that his faculties might be radically incompetent.

## Addenda.

(1) As an additional example of the mischief which comes of not viewing Descartes' words in their context, and every philosopher's words in their context, it is instructive to observe how falsely St. Augustine has been quoted as a precursor of Descartes. St. Augustine does indeed use the very valid argument, that the existence of self is invincibly brought home to the conscious individual, and that it is asserted even in the act of doubting. But St. Augustine does not preface the argument by a suicidal declaration of scepticism, nor does he fall into the vicious circle of proving reason from God, and God from reason. Without first taking himself the fatal cathartic of universal doubt, but arguing against the possibility of universal

doubt, he has passages like these: "If a man doubts, he lives; if he doubts that he doubts, he understands. If he doubts, it is because he wants to be certain. If he doubts, he thinks. If he doubts, he is conscious of his ignorance. If he doubts, he deems that he ought not to assent, save on reasonable grounds."[1] "You who wish for a knowledge of yourself, do you know your own existence?" "Yes, I do." "How do you know it?" "That I don't know." "Do you know whether you are simple or complex?" "No." "Do you know that you have the power of motion?" "No." "Do you know that you are capable of thought?" "Yes."[2] Finally, "Without any delusive phantasm of the imagination, I am certain that I am, that I know and love. As regards these truths, I have no fear of the arguments of the Academics who may object: but what if you are deceived? If I am deceived, I am."[3]

Not one of the quotations sanctions universal scepticism as a prelude to philosophic certainty.

(2) By the side of Descartes' theory it is interesting to place the view of Cousin,[4] that the possible forms of philosophy are four, *sensism* and *idealism*, each leading to *scepticism*, which in turn has for its reaction *mysticism*. He denies that scepticism can come first, being necessarily preceded by dogmatism, either sensistic or

---

[1] "Si dubitat, vivit. Si dubitat, dubitare se intelligit. Si dubitat, certus esse vult. Si dubitat, cogitat. Si dubitat, scit se nescire. Si dubitat, judicat se non temere consentire oportere." (*De Trinitate*, 14.)

[2] "Tu, qui vis te nosse, scis esse te? Scio. Unde scis? Nescio. Simplicem te scis, an multiplicem? Nescio. Movere te scis? Nescio. Cogitare te scis? Scio." (*Soliloq.*, Lib. II. cap. i.)

[3] "Sine ulla phantasiarum et phantasmatum imaginatione ludificatoria, mihi esse me, idque nosse et amare certissimum est. Nulla in his vereor. Academicorum argumenta formido, dicentium, quid si falleris? Si fallor sum." (*De Civ.*, Lib. XI. c. 26.)

[4] *Histoire de la Philosophie*, Leçon 13me.

idealistic. "Negation is not the starting-point of the human mind, as it pre-supposes that there is something to be denied, hence something that has previously been affirmed. Affirmation is the first act of thought. Man, therefore, begins with belief, belief in this or that; and so the first system is dogmatic. Its dogmatism is either sensist, or idealistic according as the thinker trusts respectively thought or the experience of the senses. Mysticism marks the despair of the human mind, when after having naturally believed in itself, and started with dogmatism, it takes refuge from scepticism in pure contemplation, and the immediate intuition of God. Such is the necessary sequence of systems of thought in the human mind."[5]

(3) Descartes is a warning against over-confidence in self for the working out of a new system. He complained that philosophy presented the appearance of a city built by many hands at different times; and he argued that a symmetrical whole required unity of workmanship. He tried himself to be the single workman, who should build up the whole of an enormous city, after first pulling down the old structures; but in both respects his efforts were failures, monumental failures for the warning of posterity. In a matter so open to human thought as the nature of its own

[5] "L'esprit humain ne débute pas par la négation; car, pour nier, il faut avoir quelque chose à nier, il faut avoir affirmé, et l'affirmation, c'est le premier acte de la pensée. L'homme commence donc par croire : il croit soit à ceci, soit à cela, et le premier système est le dogmatisme. Ce dogmatisme est sensualiste ou idéaliste, selon que l'homme se fie davantage ou à la pensée ou a la sensibilité. Le mysticisme, c'est le coup de désespoir de la raison humaine, qui après avoir cru naturellement à elle-même et débuté par le dogmatisme, effrayée par le scepticisme, se réfugie dans la pure contemplation et l'intuition immédiate de Dieu. Tel est l'ordre nécessaire du développement des systèmes dans l'esprit humain." (*Ib.*)

certitude, no man of proper modesty should venture upon the boast; Heretofore the world has gone wrong, but at last *ecce ego!* Even the gentle Ferrier ventures to claim a few of these downright new discoveries; but they are, of course, all delusions: and of Comte, who ceased to read other philosophers in order to develop his own thought, Mill says that he developed "a colossal self-conceit."

# CHAPTER X.

## THE PRIMARY FACTS AND PRINCIPLES OF THE LOGICIAN.

*Synopsis.*
1. The philosopher's mental outfit in general when he starts on his course.
2. The disengagement of certain great primaries, notwithstanding the complicated condition of adult thought, and the impossibility of reverting to the first thoughts of childhood.
   (*a*) The primary fact in all knowledge. (*b*) The primary condition of all knowledge. (*c*) The primary principle of all knowledge.
3. Other primaries may be asserted, but the above three deserve special mention.

*Addenda.*

1. The outfit as to bodily means, with which some begin a University career, has excited partly the amusement and partly the compassion of those who have heard such stories as are typified, on one side by the youth with the "great coat and the pair of pistols;" and on the other side by some of the poorer students of Glasgow and Edinburgh, who all too grimly appreciate Sydney Smith's joke: "We tune our song on slender oats."[1] Still some manage to feed fat the mind, while the flesh remains lean, especially if they start with a

[1] " Tenui musam meditamur avena." (Virgil, *Ecl.* i. 2.)

good mental outfit: for that is the immediately important thing in the freshman. The philosopher's stock-in-trade at starting, after the clearing out of his premises, has been reduced by Descartes to what we have seen to be ruinous conditions: and therefore we naturally ask ourselves with what supplies we undertake to make a commencement. Already we have settled to keep our natural knowledge, not in the extravagant trust that all our judgments have been correct, but with a general assurance that we have fairly trained minds, and have laid in a store of certitudes, the ultimate foundation of which we may proceed to examine at leisure, without the slightest fear of bringing about a total collapse. We did not begin systematically to philosophize during our school life, because we were not ripe for the exercise; but we began in early manhood, when at least we might hope that we were moderately prepared for the work. We should have held it preposterous had we been called upon, at the inaugural lecture of our philosophic course, to recite, instead of a *Credo* a *Dubito*, after the style of the Cartesian formula: I doubt all the truths which hitherto I have held most certain; I question the reality of my body, and the reports of all my senses; I doubt the competency even of my mental powers, and by means of this doubt do I expect salvation.

2. Not, however, to rest content with declaring a general trust in the results of our previous life, subject to many such accidental corrections as a more critical study of details shall suggest, we

must pick out a few primary truths, as of universal prevalence throughout every act of knowledge. It has before been declared that we cannot give, in perfect order, first a single principle, then another, and then another, and lay it down, that this is the progress, step by step, of every human mind. Much has been said, both in prose and in verse, about the first waking up of the child to conscious life, and especially to the distinction of self and not self. One sage regards the latter crisis as very solemn, and tells how the infant mind, seeing itself opposed to a whole universe, with a strong cry proclaims its right to assert its own individuality, and to live. Ferrier[2] describes the moment as one of transition from the "feral" to the "human" state. Other authors have carefully chronicled the indications of dawning intelligence in young children, and the study of new-born animals has not been neglected. Richter fancied that he remembered the time and the circumstances, in which the thought first flashed upon him, "I am I;" and he gives a detailed account of the grand revelation.

But these are matters we may leave to other inquirers. Probably anything like the clear, steady possession of one definite certitude does not come till after the mind has acquired many floating ideas, which appear and disappear fluctuatingly on the surface of consciousness, and after many judgments of similarly fluctuating character. That the child's first thoughts are fixed, clear-cut, and coherent judgments, is more than we can believe.

[2] *The Philosophy of Consciousness*, Part V. c. iii. and per totum.

Much as we dissent from the whole theory upon the origin and the nature of knowledge, as propounded by Mr. Spencer,[3] we may take some useful hints from a passage like the following: "Every thought involves a whole system of thoughts, and ceases to exist if severed from its various correlatives. As we cannot isolate a single organ of a living body, and deal with it as though it had a life independent of the rest; so, from the organized structure of our cognitions, we cannot cut one, and proceed as though it had survived the separation. Overlooking this all-important truth, however, speculators have habitually set out with some professedly simple datum or data; have supposed themselves to assume nothing beyond this datum or these data; and have thereupon proceeded to prove or disprove propositions which were, by implication, already unconsciously asserted along with that which was consciously asserted." Our own application of the doctrine will appear in what we are now to explain.

Probably it is our common experience, that we cannot, by memory, recall how knowledge first sprang up in the mind, but we can do something suggestive on the subject. We can actually remember how, upon our beginning some new study, the terms and principles one moment seemed to show a gleam of light, and then were suddenly dark again; then once more the flame flickered up, till gradually a few strong lights were fixed, around which we could range others.

[3] *First Principles*, Part II. c. ii. § 39.

And if this was the case in later years, yet more strongly would the like features be marked, when our intelligence was first feeling its way to the exercise of its own powers. The child's mind is full of abortive ideas, incoherences, and fantastic combinations; so that nurses, in talking to children, by a sort of instinctive sympathy, talk nonsense, while nonsense verses form the child's earliest literature. Some of our recollections of childhood are probably of grotesque, impossible events, which yet we should simply say that we remembered, were it not that we now perceive such incidents to be absurd as realities; they are incidents like those of the nursery rhymes, one writer of which, Mr. Lear, has had positively to defend himself against symbol-scenting interpreters, by the declaration, "nonsense plain and absolute has been my aim throughout."

So far, however, as we did form any judgment, we must have been in practical possession of certain great general principles, though we could not single out the abstract elements from their concrete embodiments and universalize them. Now at length we are called upon to evolve what must have been involved in our earliest cognitions, whatever may have been their concrete matter; nor must we overlook the difficulties in the way of our analysis. We have to abstract first principles, not out of our first thoughts, which are equivalently lost to us, but out of our adult thoughts, which are often so complicated that a single sentence may suppose an acquaintance with a vast subject-matter. Without falling into the exaggerated doctrine of relativity, we must allow those facts of which it is the perverted account, for instance, that all knowledge is closely interrelated. Reverting to the passage just now quoted from Mr. Spencer, we must allow the almost illimitable blending of idea with idea, in the texture of mind: indeed the body of our knowledge is a sort of organism, the property of which is, that the parts exist for the whole, and the whole for the parts. It will be a test that we are able sufficiently to isolate by reflexion a few primary truths, which can be absent from no act of knowledge. We insist much on this power of reflective abstraction, and by its means we are going to work. A primary fact, a primary condition, and a primary principle—these are what we are about to single out.

(*a*) The fact of his own existence is given implicitly, in every act of genuine knowledge which a man elicits. For knowledge is of no avail unless it comes home to the subject as his own; or, according to one phraseology, perception is useless without apperception, whereby the object known is, for each one, brought under the form, "I know." *Ego Cogito*, not *Est Cogitatio*, is what Descartes rightly regards as an important recognition, made by every human mind when it comes to the proper use of its powers.

An ordinary man would hardly raise any difficulty against what has just been asserted, unless he laboured under some delusion as to the extent of the assertion; fancying, for instance, that it required a clear, explicit thought about self, or a cognition of self which should amount to a definition of personality or of selfhood. To guard against such misconceptions, be it understood that the recognition of self need be only implicit, and need be no more scientific than what comes within the competency of the newly dawned reason of the child. But here precisely we are taken up. Does not a child show that it has no perception of self, by speaking of itself as "baby," "Georgie," "Maggie," in the third person? This fact proves nothing, for it is natural enough that a child should call itself by that name by which it hears others call it, instead of at once seizing upon the use of the first personal pronoun. Also there is no difficulty in allowing that self-consciousness is not as strong in the child as in the adult: and hence the simplicity and candour of children. The assertion of this characteristic is not invalidated by the counter-assertion, that there is to be met with in children the unpleasing trait of great selfishness, imperiousness, vanity, jealousy of rivals, which manifestations cannot all be shown to proceed from a sort of mere animal instinct, devoid of all intelligent perception.

If next we consider the opposition that is likely to be made against our *First Fact* from the part of philosophic theory, then the antagonism is greater than what was offered by the ordinary thinker. Still in the presence of a plain testimony of experience we have a right to disregard the mere exigence of a philosopher's system, which otherwise we know to be wrong. It is enough therefore to mention, without taking the trouble to refute, the view of Mr.

Spencer.[4] Driven by his theory to hold that subject can never be object, and that reflexion is never made upon a present state of mind but always on a past, he says that though we have a "certainty" of self, we cannot have a "knowledge of self." "The personality of which each one is conscious, and of which the existence is to each a fact beyond all others the most *certain*, is yet a thing which cannot be truly *known* at all, knowledge of it being forbidden by the very nature of thought." It is far better to assert simply, on the strength of evident experience, that we know self, than thus recur to a distinction, which supposes "a fact most certain" not to come under "knowledge," but only under some obscurer form of consciousness. If such consciousness does not amount to knowledge, it can be only a sort of blind belief; a consequence we may deduce from many other parts of Mr. Spencer's philosophy. In reliance on his principle[5] that "the invariable persistence of a belief is our sole warrant for any truth of immediate consciousness and of demonstration," he makes the unsatisfactory announcement, that "in the proposition, *I am*, he who utters it cannot find any proof but the invariable persistence of the belief in it." It is far simpler and truer to say, that to each sane man his own existence is self-evident, and admits of no strict proof; his constant belief in it not being so much a proof, as something which requires no justification by proof, because the thing is self-evident, and therefore above proof strictly so called.

[4] *First Principles*, Part I. c. iii. § 20.   [5] *Ibid*. c. iv. § 26 in fine.

(*b*) Descartes, with us so far, now abandons us, declaring that he can and does doubt the validity of his very faculties; and that in consequence he is driven to set about a scientific verification of his mental powers. We maintain that our ability to know cannot be to us matter of strict demonstration, of inference from premisses more evident, but must be taken as the *First Condition*. This is no assumption, in the bad sense of the phrase; for we are made immediately conscious of our power to know, in the very exercise of our faculties. Nor could we learn the fact any other way, as, for example, by the testimony of others. If a rational being uses his reason, the result is that he finds out what manner of being he is; a thing that the irrational being never does, especially if it be also insensate, like a plant or a stone. As Cardinal Newman puts it, we trust first of all, not our faculties, but their acts, or our faculties in act. And Dr. M'Cosh, in his *Intuitions of the Mind*, says: "We do not found knowledge, as the Scotch metaphysicians seem to do, on belief in our nature and constitution. It would be as near the truth to say, that we believe our constitution because it makes known realities. But the truth is that the two seem involved one in the other. In our cognitions and feelings, we know and believe in objects, and in doing so we trust in our constitution."

One little allowance, however, may be made to those who teach that we *prove* our ability to know, though, it is to be feared, they will not be satisfied with the concession. We must remember here what we stated in our last chapter, how a man waking slowly from a vivid dream, may gradually explore his own state and so convince himself by degrees that he is in his right mind. But such a case lends no support to the adversaries of what here is being assumed as the First Condition of all knowledge, a condition the fulfilment of which is tacitly recognised in every intelligent act that we perform. "Knowledge is power," and feels itself to be such intrinsically: it feels that it is a power to know.

(c) Within the thinking subject we have now got a *First Fact*, the recognition by the subject of self; and a *First Condition*, the subject's power to know, also recognised as a fact; we must next add *a First Principle* on the objective side, namely, the *Principle of Contradiction*. To show the objectivity of this principle we formulate it, not on the logical, but on the ontological side. We do not simply say, "the same thing cannot, in the same sense, be affirmed and denied," but "the same thing cannot, in the same way, be and not be." Under both aspects the principle is self-evident, and it is only the extreme of irrationality in Mill, which makes him refrain from asserting its absoluteness both for all thought and for all things. Yet even he ventures so far as to write,[6] "that the same thing should at once be and not be; that identically the same statement should be both true and false, is not only inconceivable to us, but we cannot conceive that it should be made conceivable." He admits too, that if there are any primitive necessities of thought, this is one of them. With him Mr. Bain agrees to the extent of affirming

[6] *Examination*, c. vi. p. 67; cf. c. xxi. p. 417. (2nd Ed.).

that, "were it admissible that a thing could be and not be, our faculties would be stultified and rendered nugatory." Hampered by no theories from Hume, we simply assert, as self-evident to reason, the Principle of Contradiction, or as Hamilton prefers to call it, the Principle of Non-contradiction. No statement that we could make would have any meaning, if this principle had not clear objective validity.

3. The above three are called *primaries*, but not in the exclusive sense. Such a phrase as the "three first" is often criticized, and by some declared to be quite inadmissible. If it stands for objects which are respectively first, second, and third in a series, we may leave it undiscussed; but when it stands for three which are abreast in forming the first rank, then we are here concerned to defend the expression, so far as to justify our assertion of "three primaries." The word "first," like any superlative, may qualify simply an individual, or it may qualify a whole class, and be predicated of the individuals in that class. Thus we can use it when we say, "the ten first men in England:" each of the ten holds independently a first place. When, therefore, we are speaking of the three primaries, we are not putting one before the other, nor even denying that there are other primaries: it is sufficient that the three are primaries, and further, that among primaries, they deserve a special prominence to be given to them, because of their importance. But, in addition to them, the principle of identity is primary, so is the principle of sufficient reason, that nothing can be without an adequate

account for its existence ; and so is the principle of evidence, that what is evident must be accepted as true. To compile a catalogue of all the truths which are self-evident, and cannot be reduced to components simpler than themselves, would be a tedious work, and not helpful to present purposes. If, however, we are called upon to emphasize any beyond the three mentioned primaries, it will be the Principle of Sufficient Reason, so often violated by pure empiricists, and yet so vital to all philosophy. When Mr. Bain declares that there is no repugnancy in "an isolated event," or "in something arising out of nothing," if we are to take him literally, he puts himself out of the pale of reasoning creatures. His friend Mill is nearer to the sane principle, at least as far as a single sentence goes, when he writes: "That any given effect is only necessary provided that the causes tending to produce it are not controlled ; that whatever happens could not have happened otherwise, unless something had taken place, which was capable of preventing it, no one needs surely to hesitate to admit." Unfortunately when he says, "cause," Mill does not mean "cause," but otherwise his words are in the right direction ; and we at any rate do well to put in the position of a primary truth, the principle of Sufficient Reason.

We must dissent, however, from the peculiar treatment of this principle by Mansel, who first of all states it only in its logical side, "Every judgment must have a sufficient ground for its assertion," and then denies it to be a principle. "The only reason for a thought of any kind is its relation to some other thought, and this relation will in each case be determined by its own proper law. The principle of sufficient reason is, therefore, no law of thought, but only the statement that every act of thought must be governed by some law or other."[7] He even ventures something like a possible suspicion of the principle, but does not clearly assert it: "If considerations [concerning free will] suggest a limit to the universality of the principle of sufficient reason, so be it."[8]

### ADDENDA.

(1) Mill[1] declares "there is no ground for believing that the *Ego* is an original presentation of consciousness." When it does become such we have the following account of it: "The fact of recognising a sensation of remembering that it has been felt before, is the simplest and most elementary fact of memory; and the inexplicable tie, or law, or organic union, which connects the present consciousness with the past one, of which it reminds me, is as near, I think, as we can get to a positive conception of self. That there is something real in this tie, real as the sensations themselves, and not a mere product of the laws of thought, without any fact corresponding to it, I hold to be indubitable. . . . Whether we are directly conscious of it in the act of remembrance, as we are conscious in fact of having successive sensations, or whether according to the opinion of Kant we are not conscious of self at all, but are compelled to assume it as a necessary condition of

[7] *Proleg. Log.*, c. vi. pp. 198, 223.
[8] C. v. p. 153.
[1] *Examination*, Appendix, p. 256. Compare the Appendix to Hume's *Treatise*, at the end of Bk. I. Part IV. p. 559.

memory, I do not undertake to decide. But this original element which has no community of nature with any of the things answering to our names, and to which we cannot give any name but its own peculiar one, without implying some false or ungrounded theory, is the *Ego* or Self. As such I ascribe a reality to the *Ego*—to my own mind—different from that real existence as a Permanent Possibility, which is the only reality I acknowledge in matter."

(2) There have been authors, whose connexions may be traced back at least as far as Heraclitus, and who, under the idea of "becoming," as distinguished from "being," try to do away with the asserted contradiction between simultaneous being and not being. Ferrier[2] explains Heraclitus thus: "When he says that all things are in a continual state of flux, that a thing agrees with itself and yet differs from itself; when he says that strife is the father of all things, that everything is its own opposite and both is and is not, he means that things are continually changing, or that the whole system of the universe is a never-ceasing process of 'becoming.'" "The principal feature in the conception of 'being' is rest, fixedness. Now the opposite of this is the principal feature in the conception of 'becoming.' It is unrest, unfixedness. A thing never rests at all in any of the changing states into which it is thrown. It is in that state and out of it in a shorter time than any calculus can measure."

The fallacy often used to illustrate this theory, is to suppose that mere unextended points of time and space are, not merely limits, or ideal boundaries marking divisions of time and space, but are their actually constituent elements; so that extension is made up of an infinite row of inextensibles placed side by side. This

[2] *History of Greek Philosophy; Remains*, Vol. I. pp. 114, 116.

notion is absurd, and is not held even in what is known as the dynamite theory of matter which asserts at least extended areas of force, the centres only of which are unextended points. But observing the fallacy, let us see how it is worked. A body, moving continuously, is supposed at once to arrive at any given point, and to leave it at the same moment, and thus to be at once there and not there. The sophism lies in making the point to be at once part of the line and not part of its extension. If we keep to definitions, a point of time is of no duration, and a point of space of no extent. When, then, we say that a body moves over a point of space in a point of time, we are uttering the very true statement, that in no time no space is traversed. It being clear, therefore, that to account for the traversing of a literal point in a body's path is to account for no part of the path at all; it is equally clear that if any part is to be accounted for, then we must take at least some small extent both of space and of time. But as soon as extension is considered, the whole argument fails: it can no longer be pretended, that the body together is and is not at one place.

A somewhat like fallacy is used in reference to circular motion, which may be considered to be composed of a projection in a straight direction and a constant attraction by a definite law to the centre. The result is that the body never gets either nearer to the centre or further from it, the curvilinear path is the compromise between the two motions, but it is never one component alone. Here steps in the fallacy-framer, and pretends that the motion is both tangential, away from the centre, and centripetal, or towards the point of attraction. We answer firmly, there is no such union of contradictories, there is only a movement of revolution, which is never for a moment either centrifugal or centripetal.

(3) Mill's empirical account of the induction by which we reach the principle of contradiction, is thus given: " The principle of Contradiction should put off the ambitious phraseology which gives it the air of a fundamental antithesis pervading nature, and should be enunciated in the simpler form, that the same proposition cannot at the same time be false and true. But I can go no further with the Nominalists, for I cannot look upon this last as a merely verbal proposition. I consider it to be, like other axioms, one of our first and most familiar generalizations from experience. The original foundation of it I take to be, that Belief and Disbelief are two different mental states, excluding one another. This we know by the simplest observation of our own minds. And, if we carry our observation outwards, we also find that light and darkness, sound and silence, motion and quiescence, equality and inequality, preceding and following, succession and simultaneousness, any positive phenomenon whatever and its negative, are distinct phenomena, pointedly contrasted. I consider the maxim in question to be a generalization from all these facts." [3]

(4) There is a limit to human patience in bearing with subtleties, which have for their object the overturning of such fundamental principles as that of contradiction; and in illustration of the way in which exhausted patience rebels, a few examples may be borrowed from Janet's little book on *Materialism*.

Hegel's dialectic process, which goes on the theory of reconciling contradictories by successive steps of antithesis and synthesis, was allowed a certain degree of triumph; but it also called forth violent denunciations from its opponents, and led to wide divergencies between its friends. Schopenhauer expressed a common

[3] *Logic*, Bk. II. c. vii. § 4.

feeling when he called such philosophy "a minimum of thought, diluted into five hundred pages of nauseous phraseology." Humbolt, accustomed to the more sober physical sciences, turned to ridicule what he called "the dialectic tricks" of Hegel; while Goethe avowed that, " if the transcendentalists ever became aware of it, they would find themselves to be very absurd."

As a reaction against so much idea-weaving, and so much building up in the clouds, there arose the gross materialism of Moleschott, Büchner, and Vogt. The second of this trio pronounced the pretended philosophy to be "verbiage," "jargon," "metaphysical quackery," "a cooking up of old vegetables under new names," and a proceeding "which inspires legitimate disgust in learned and unlearned alike."

(5) Hardly as a serious objection to the principle of contradiction, and yet as furnishing a straw at which a desperate opponent might clutch, but still more as having an interest of its own, the fact may be mentioned, that of late years lists have been compiled of words from out-of-the-way languages, which have a double signification, namely, an idea and its opposite. We are not quite without examples of the kind in more familiar tongues. The case illustrates, so far as the saying is true, the old *dictum*, that "the knowledge of opposites is one." Another observed fact of an analogous order is that people recovering from amnesia, or loss of memory, are found using, instead of the right word for a conception, just its opposite. To these or any other similar discoveries the friends of the Hegelian identification of contradictories are welcome; but their cause will remain hopeless as ever.

(6) At the root of much difficulty made against the isolation of primary, absolute principles, stands the theory of Relativity in all knowledge, on the strength of

which the notion of absolute being is denied to us; and what is refused us under the title of "knowledge," at last is given back to us under the name of an inferior mode of consciousness. A sentence omitted in a quotation lately made from Mr. Spencer, shall here be supplied:[4] "The development of formless protoplasm into an embryo, is a specialisation of parts, the distinctness of which increases only as fast as their combination increases—each becomes a distinguishable organ, only on condition that it is bound up with others, which have simultaneously become distinguishable organs: and similarly, *from the unformed material of consciousness*, a developed intelligence can arise only by a process which, in making things definite, also makes them mutually dependent—establishes among them certain vital connexions, the destruction of which causes instant death of the thoughts." Now if we refer back a little, we shall learn something about what this "unformed material of consciousness" is supposed to be.[5] "We come face to face with the ultimate difficulty—how can there possibly be constituted a consciousness of the unformed and unlimited, when by its very nature consciousness is possible only under forms and limits? In each consciousness there is an element which persists. It is alike impossible for this element to be absent from consciousness, and for it to be present in consciousness alone; either alternative involves unconsciousness—the one from want of substance, the other from want of form. But the persistence of this element under successive conditions, *necessitates* a sense of it as distinguished from the conditions. The sense of this something, conditioned in every thought, is constituted by combining successive concepts deprived of their limits and con-

---

[4] *First Principles*, Part II. c. ii. § 39.
[5] *Ibid.* Part I. c. iv. § 26, p. 94.

ditions. The indefinite concept is not the abstract of any one group of ideas, but of all ideas, namely, EXISTENCE, which is an indefinite consciousness of something constant under all modes. Our consciousness of the unconditioned being literally the unconditioned consciousness, or *raw material* of thought to which in thinking we give definite forms, it follows that an ever present sense of real existence is the very basis of our intelligence. At the same time that by the laws of thought we are rigorously prevented from forming a conception of absolute existence, we are by the laws of thought equally prevented from ridding ourselves of the *consciousness* of absolute existence; this consciousness being the obverse of our own self-consciousness. And since the only possible measure of relative validity among our beliefs, is the degree of their persistence in opposition to the efforts made to change them; it follows that this which persists at all times, under all circumstances, has the highest validity of any." In brief, our highest *belief* is about a matter we cannot *know;* but about which we have an *indefinite consciousness*.

'riminaries — { First Fact. Recognition by the subject of self.
"three first" { — Condition. subject's power to know.
.. Principle. That of Contradiction — i.e. the same thing cannot, in the same way, be and not be.

Principle of Identity.
.. .. Sufficient Reason. Nothing can be without an adequate account for its existence.
— .. Evidence. What is evident must be accepted as true.

# CHAPTER XI.

### RETROSPECT AND PROSPECT.

*Synopsis.*
1. Retrospect.
2. Prospect.

THE last proposition has brought us to a point whence a look backwards, and another forwards, become necessary in order to clear away natural misgivings that we may be wandering about aimlessly. We have travelled together through regions of our own experience as knowledge-gathering creatures; we have noted down the general characteristics of certitude and of its allied or opposed states, but have avoided details. The consequence may be that some of the company have felt uneasy, and would over and over again have liked to pause on some such questions as, how the reports of the senses are to be credited, or how abstract and general ideas are valid, which confessedly have corresponding to them no abstract and general objects. But steadily and inexorably the surveying party has been led on, with the promise that another survey shall be made to fill in details, and with the declaration that, meanwhile, human certitude, before our philosophizing about it, sufficiently attests its own validity.

1. We have mapped out some of the general features of human knowledge, and spreading out the unfinished sketch, we observe what we have done. Beginning with logical truth, that is, with the knowing of truth, we decided, that apart from any theory as to how the mind can produce a resemblance of the several objects which it knows, yet we cannot intelligibly admit that it really knows anything while we deny that the knowledge bears any likeness to the thing known. Some sort of likeness there must be, though after a peculiar mode which our imitative arts cannot copy. Mere concomitant variation in mind and object will not suffice, if it is declared to carry no resemblance.

Inquiring next what is the special act of mind in which logical truth is to be found in its fulness, we settled that it must be the *judgment*, the act by which we affirm or deny, by which we are conscious that something is, or is not. Unless we go as far as this point, we are not yet in possession of a truth; at best we are on the way to possession.

The conscious, full, and firm possession of the truth, to the exclusion of doubt, is *certitude*, a state of mind which we contrasted with ignorance, and with mere tendencies to assent, or assents given as to probabilities only. To distinguish these states belongs to the logician, though it is not his province to determine, in all fields of knowledge, what is the measure of assent or dissent due to any given statement. As a matter of self-analysis, a man may sometimes be puzzled whether or not he ought to put aside suggested reasons for doubt, as being quite neutralised by contrary reasons; and in cases of such perplexity he will often have to appeal to considerations more concrete than logic supplies.

Returning to certitude we gave its broad distinction into *natural* and *artificial, non-scientific* and *scientific, philosophic* and *common-sense;* and we showed the interdependence between the two. Either branch—but we have regard especially to the second—is divisible according to its specific motive, into three kinds, *metaphysical, physical,* and *moral.* We likewise saw in what sense a proposition, which is certain, may be regarded as having its certitude greater or less.

In absolute opposition to certitude came *scepticism* under its most uncompromising form, or total negation of the power of mind to acquire real knowledge of things. Such scepticism was shown to be quite indefensible as a position taken up and defended by argument; its very possibility was denied in view of the irresistible self-assertion of a reasonable nature. However, there was a scepticism calling itself *methodic,* and professing to lead to the most legitimate dogmatism; but its professions proved hollow, and its failure served only to confirm our own previous proposition, that philosophy must build on natural certitude. In the words of Mr. Spencer, the philosophy of certitude " can be nothing but the analysis of our knowledge by means of our knowledge, an inquiry by our intelligence into the decisions of our intelligence." We cannot carry on such an inquiry without taking for granted the trustworthiness of our intelligence. But against any one supposing that this assumption itself is a blind, instinctive process, we entered our "caveat" not without call.

Having rejected the Cartesian primary facts and principles, as explained by their author, we felt bound to agree upon some of our own; and as primary truths we assigned what were called the *First Fact*, the *First Condition*, the *First Principle;* to which trio the *Principle of Sufficient Reason* was added. Out of these elements we cannot hope to build up a system as Euclid built up his geometry; but so far as the logic of certitude is reducible to a few elements, these are they. We need hardly try to make all that Hamilton has made out of the Principle of Identity; because so far as what he says has truth in it, the truth seems scarce worth such explicit proclamation; or at any rate, it is very calculated to vex the souls of some readers. In behalf of our own primaries, the defence is available, that they are evident without demonstration, and that no one can argue against them without implicitly affirming them.

2. Thus far we have gone; but what is to be the next step? Many schoolmen follow the plan of entering here upon the consideration of what they call the means or the sources of knowledge. Their work comes pretty much to a division and a defence of faculties which successively take up the elements of knowledge, and bring them out in the shape of formed propositions. A justification is attempted of

sensations, ideas, memory, judgment, and reasoning. But without a word of condemnation for the method of others, we may relegate these matters to the Second Part; the reason being that they may fairly be regarded as belonging to the details of the Subject, not to that most general description of Certitude which forms the First Part. As belonging to the latter, however, we will at once grapple with a question often delayed till the very end of the treatise, namely, with *Evidence*, considered as the objective criterion of truth. Since this is the perfectly general criterion of all certitude, we are justified in putting it along with the other matters which we have called "Generalities." There will thus be a book on Generalities and a book on Particularities; after which the reader will not be asked to extend his patient efforts to yet another book.

# CHAPTER XII.

## THE REJECTION OF VARIOUS THEORIES ABOUT THE ULTIMATE CRITERION OF CERTITUDE.

*Synopsis:*
1. Blind impulse to believe.
2. Verification by the senses.
3. Traditionalism.
4. Some sort of vision of things in God, or in divinely communicated ideas.
5. Clear and distinct ideas as asserted by Descartes.
6. Consistency.
7. Inconceivability of the opposite.
8. Concluding remarks.

As builders clear the ground before they begin to build, so we shall do well to start by putting out of the way certain proposed criteria of truth, which either we cannot accept as criteria at all, or else not as ultimate criteria.

1. Some philosophers, often more in appearance than in reality, or more as an occasional aberration than as an opinion steadily maintained throughout, represent the cause of our assents to be, in last analysis, a *blind instinct to believe*. What is true in their doctrine is, that we cannot penetrate the secret of the intellectual act, and see how it is that this most wonderful act, the act of knowledge, is elicited from the faculty. The conscious process

we are aware of because it is conscious; but the physical process, so to term it, we do not comprehend. When we think of the marvellousness of intelligence, we are quite lost in the mystery of the process, and almost feel inclined to doubt whether our knowledge is not illusion. To this extent intelligence gives no explanation of itself. But to say that we assent by a blind instinct, is to take out of the assent its percipient character, to render it non-intellectual, to make it a contradiction in terms. Allowing, therefore, that the manner in which we understand is impenetrably dark, we cannot allow that the understanding itself acts in the dark, by means of blind instinct. Its essence is to see its way as it goes.

2. The first proposal can hardly be called that of a criterion, for a criterion supposes something genuinely intellectual; but the second proposal does offer something which, at least, is in the cognitive order, though in the lowest grade of cognition. The criterion is *verification by the senses*. Lewes, who, in his *Problems of Life and Mind*, is one of its strong advocates, insists that the great mass of our thoughts, being abstract, generalized products, are only symbolic of the real, and must be reduced to their first origin in sensation, if their value is to be tested. Our sensations are as the arithmetic of objects, our conceptions are as the algebra, that is, symbolic expressions. Besides the criterion of sense, however, he allows a secondary, derivative criterion, which consists in reduction to intellectual intuition.

Mill cannot quite be put in the same class with Lewes, for he speaks of the necessity we are under to accept all averments of consciousness, provided that they can be shown to belong to its pure, primitive state. Still the following passage will show how inclined he was to make sensation a sort of ultimate test: "When I say that I am convinced there are icebergs in the Arctic sea, I mean that the evidence is equal to that of my senses; I am as certain of the fact as if I had seen it. And on a more complete analysis, when I say that I am convinced of it, what I am convinced of is, that if I were on the arctic seas I should see it. We mean by knowledge and by certainty an assurance similar and equal to that effected by our senses. If the evidence can in any case be brought up to this, we desire no more."[1]

Here Mill evidently is speaking, not of mere sensation, but of intellectual perceptions following after sensations. However, the precise nature of neither his doctrine nor of that of Lewes need trouble us at present; for we want no accurate estimates of different philosophies, but only a refutation of the broad proposition, that the ultimate criterion of truth is verification by the senses. Now a sufficient objection to this view is the two-fold fact, that a mere sensation, as such, cannot be the direct criterion for an intellectual faculty, and that we have many certitudes about objects which are supra-sensible. What, however, we may allow to verification by the senses is, that often a physical

[1] *Examination*, c. ix. in initio.

theory, carried through several steps by the mere reasoning process, requires to be brought to the test of observation or experiment, in order to make sure that the reasoning is consecutive and leaves out none of the involved data. Thus it was right to look actually with the telescope for the planet, the position of which Adams and Leverrier had mathematically calculated. But in all cases alike certitude itself is intellectual, and must have a criterion directly intellectual.

3. Distrustful of self, man is inclined to make his last appeal to his fellows, especially to the majority of men; and more especially to the majority, if they are supposed to be the divinely appointed custodians of a primitive revelation. Thus we have the *appeal to Tradition* as an ultimate criterion of truth. *Traditionalism* is a doctrine which has had some vogue in France. Long ago our own John of Salisbury had written: "As both the senses and human reason frequently go astray, God has laid in faith the first foundation for the knowledge of truth."[2] A sober interpretation may be given to a sentence like this, but Bayle was outraging alike God and man, when he pretended utterly to discredit human reason, in order to make way for the sole reign of faith. "Human reason is a principle of destruction, not of construction; it is capable solely of raising questions, and of doubling about to make a controversy endless. The best use that can

---

2. " Quia tum sensus quum ratio humana frequenter errant, ad intelligentiam veritatis primum fundamentum locavit Deus in fide." (*Metalogicus*, Lib. IV. cap. xiii.)

be made of philosophy is to acknowledge that it can but set us astray, and that we must seek another guide, which is the light of Revelation."[3]

In recent times the principle here enunciated has been taken up by men far more earnest than Bayle, but all their earnestness has failed to make a dangerous doctrine safe. The pith of De Bonald's teaching is given in a single sentence of his: "This . . . proposition, Thought can be known but by its expression, that is, by speech, sums up the whole science of man."[4] Taking up the idea of De Bonald, De Lamennais, in his famous *Essai sur l'Indifférence dans la Matière de Réligion*, elaborated a scheme of traditionalism. He supposed a primitive communication of truth from above to the race. Then, working on a principle which Aristotle mentions but does not sanction, and which Lord Herbert of Cherbury, in his treatise *De Veritate*, had adopted, namely, "what appears to all men, that is true," he embraced it to the extent of affirming that the consent of the majority determines what is the authentic tradition, or, in other words, what is the truth.

A most glaring objection to the theory starts up at once in the shape of the obviously raised ques-

[3] "La raison humaine est un principe de destruction, et non pas d'édification; elle n'est pas propre qu' à former des doutes, et à se tourner à droit et à gauche pour éterniser une dispute. Le meilleur usage qu'on puisse faire de la philosophie est de connaître qu'elle est une voie d'égarement et que nous devions chercher un autre guide, qui est la lumière révélée."

[4] "Cette proposition rationelle, la pensée ne peut être connue que par son expression, ou la parole, enferme toute la science de l'homme."

tion, "If the consent of mankind is the ultimate test of truth, how do we know that such is the fact, and how do we judge, in any particular case, what is the view of the majority?" De Lamennais himself acknowledged his inability to furnish a precise reply; but all the same he adhered to his traditionalism. "The first man receives the primary truths on the testimony of God, the highest Reason. These truths are preserved for mankind, as being ever set forth by universal testimony, which is the expression of general reason, of common sense."[5] Whence he argued that the first act of intelligence is an act of faith; so necessarily, that unless a man will begin with "I believe," he will never arrive at "I know."

With a view to giving his opinion an air of reality, De Lamennais laboriously collected, from many languages, testimonies to the opinion that primitive man drew from divine sources, and that present controversies are to be settled by reference to what has been taught from the beginning. In its right place the principle of tradition is sound enough, and that right place is pre-eminently the position of the *depositum fidei*, the body of revealed truths committed by Christ to the keeping of His Church; but De Lamennais puts the principle into a wrong place altogether. It is impossible that man should ever give, as the ultimate reason of his belief, "Because I was told;" when and why he should

[5] "Le premier homme reçoit les premières vérités sur le témoignage de Dieu, raison suprême, et elles se conservent, parmi les hommes, perpétuellement manifestées par le témoignage universel, expression de la raison générale."

accept what he is told, is always a question going deeper down.

Apart from any faith in a revelation, some might urge the *consent of the majority of men* as a natural rule of truth. Against them it suffices to say that such rule, for the most part, cannot be reduced to practice, and is sometimes fallible, never ultimate. Yet there is a great truth hinted at, namely, the impossibility of any one man discovering everything for himself by independent research, without the aid of the accumulated treasures of the age. What could Newton have done, had he been born into an age when the simple rules of arithmetic formed all that was known of mathematics? An important condition of progress is, that knowledge should accumulate; and a sufficient cause of unprogressiveness in animal intelligence is its want of power properly to preserve and build upon a tradition. There is, of course, among the lower animals some sort of heredity in matter of transmitted experiences; but there is not, in the human sense, a power of tradition and development. Man has this power, and it is his wisdom not to sacrifice it by self-isolation.

4. *Blind instinct* we have rejected as being outside the pale of knowledge altogether; *verification* by the senses as being the lowest grade of cognition, so long as it means mere sensitive knowledge; *tradition* as being inadequate and never ultimate; and now we come to a pretended *vision of things in God*, or in *divinely infused ideas*, which also we must reject. The chief arguments of those who hold such opinions, run on the lines that without

Divine aid we could not have the knowledge of which we find ourselves possessed. The best mode of replying to the so-called demonstrations, is to show that they amount to no more than so many ways of re-stating the dangerous assumption, that human faculties have not the natural power of intelligence, but must, at least to a large extent, have their work done for them by their Creator. No such helplessness can be proved, and the assertion of it sounds more injurious than honourable to God. Our experience is, not that we descend from ideas or principles which are a gift, down to our own concrete applications of them, but that we ascend from concrete facts to abstract ideas and principles; nor that we travel from a knowledge of the divine to knowledge of the created, but that our course lies from the created to the divine. The fewness of the supporters of what may be called the view of Malebranche, makes it unnecessary to go at length into the two charges against it, which are that it brings no proof and goes contrary to rightly interpreted experience.[6]

5. To assert that *clear and distinct ideas* are the ultimate test of truth, might be correct if the clearness and distinctness were sufficiently shown to be more than subjective feeling, and to be founded on objective evidence. What has been explained of the system of Descartes was enough to make manifest his great shortcomings in this particular; nor does Spinoza give a more satisfactory shape to the theory when he teaches that true ideas are

[6] See Part II. c. ii. Addenda (3).

guaranteed by the consciousness of truth wherewith they are accompanied. Of course from the subjective side our certainty is our consciousness that we are certain; but the objective side also needs to be fully stated, whereas both by Descartes and Spinoza it is neglected. In the next chapter it will form the main subject of inquiry.

6. As truth can never conflict with truth, what proves inconsistent in its parts cannot, as a whole, be true. As a secondary test of truth, therefore, *consistency* is useful; but it cannot be made the ultimate criterion, for there may be consistency in error. The wider and the more varied is the range of the consistent statements, the higher, *cæteris paribus*, is the probability of their being true; still if we allow that consistency throughout our judgments is all we can produce in proof, while we can never tie down the consistent whole of our thoughts to objective reality, our ideas are still a floating mass, well compacted together, but anchored safely to nothing substantial. We may have a beautiful arch, key-stone included, but what if there are no pillars for it to rest on?

It is, therefore, lamentable to find so many writers declaring the inability of man to get anything beyond consistency as a basis of certitude. Of necessity they must speak thus who push the doctrine of relativity to extremes; but others adopt the criterion under less pressure from their system: "We cannot," says Mansel, "know what truth is in relation to a non-human intellect; and truth in man admits of no other test than the harmonious consent of all the human faculties." This must be interpreted in conformity with the principles laid down by the author,[7] that we cannot test the absolute validity of our own mental laws, but that we must trust our Creator for having given us powers sufficient for our present state of probation, and rely upon it "that the portion of knowledge of which our limited faculties are permitted to attain to here may indeed, in the eyes of a higher Intelligence, be but partial truth, but cannot be absolute falsehood. But believing this, we desert the evidence of reason to rest on that of faith; and of the principles on which reason itself depends it is obviously impossible to have any other guarantee." Thus we are left with the incomplete result "that the laws to which our faculties are subjected, though not absolutely binding on things in themselves, are binding upon our mode of contemplating them:" a conclusion which leaves us open to many of Kant's sceptical difficulties. Again, Mr. Spencer,[8] whose further test, from the inconceivability of the opposite, will be considered presently, thus expresses himself: "There is no mode of establishing any belief, except that of showing its entire congruity with the other beliefs. Debarred as we are from anything beyond the relative, truth raised to its highest form can be for us nothing more than perfect agreement, throughout the whole range of our experience, between those representations of

[7] *Prolegomena Logica*, c. iii. pp. 73—77. Compare Professor Veitch's *Institutes of Logic*, § 43, p. 29.
[8] *First Principles*, Part II. c. ii. § 40; *Psychology*, Part VII. c. i.

things which we distinguish as real. The establishment of congruity throughout the whole of our cognitions constitutes philosophy." Thus with Mr. Spencer the avowed process is to assume provisionally the simple states of consciousness; upon these to elaborate a system; and in the end to claim acceptance for it on the plea of the complete congruity which has resulted from philosophizing with the assumed elements for starting-points. Two more instances shall be borrowed from quite a different school of thought to that of Mr. Spencer. "The ultimate test of each truth," writes Mr. Caird, in his work on Kant, "a test which at the same time fixes the limit of its validity, lies in the exhibition of its relation to other truths in a system. Thus philosophy is a kind of reasoning in a circle; but this is no argument against it, for it is the circle beyond which nothing lies. The ultimate unity of knowledge must be that in which all the elements of knowledge are reflected into each other; in which the parts cannot be apprehended except as merging in the whole, and the whole cannot be apprehended except as necessarily differentiating itself into parts. The essential presupposition of all philosophy is, that the world is an intelligible system, and therefore capable of being understood and explained." This view becomes all the more intelligible if read in the light of a Hegelian principle which Mr. Wallace, at the beginning of his work on *The Logic of Hegel*, thus enunciates: "All the objects of science, all terms of knowledge, lead out of themselves, and seek for a centre and resting-point. They are severally inadequate and partial, and crave adequacy and completeness. They tend to organize themselves, and so to constitute a system or universe, and in this tendency to unity consists their truth : their untruth lies in isolation and pretended independence. This completed unity in which all things receive their entireness and become adequate is their truth: and the truth as known in religious language is God."

If consistency throughout the entire body of truths were the only criterion, even the most learned man could never make use of it, for he never knows all truths ; and the man of little education could hardly claim any certitude, for his knowledge is so limited, and he has done nothing to harmonize the different parts of his slender stock. On the other hand, as a fact, the ablest thinker among men may, on secure grounds, hold truths, the consistency of which he fails to perceive, though of course he perceives no positive inconsistency. When further we repeat that consistency alone, without a guarantee of objectivity, is insufficient, we have given reasons enough for rejecting the proposed criterion. A consistent novel is not history, and a consistent account of the evolution of the universe is not proved true till it be connected with reality. A theory like that of La Place might be possible, without being verified in fact.

Still consistency is an excellent test in its own sphere, and Mr. Spencer might have been saved some of the chapters which he has unfortunately written had he been more alive to the use of his own criterion, consistency. For example, Part I. of his *First Principles* is largely employed in drawing up a list of antinomies, which, on his theory of knowledge, are forced upon the human mind. Now these antinomies are not saved from being inconsistencies of assertion, by his adroit distinction between *knowledge* and *indefinite consciousness*. Verbiage apart, it is inconsistent to maintain that we must firmly *believe* the existence of the Absolute, but must deem it quite *unknowable;* that we must *believe* in the Non-Relative, but confine our knowledge to the Relative. Just what Mr. Spencer wants is escape from his doctrine of Relativity.

7. *Inconceivability* being itself a negative term, does not promise well, at first sight, to be a good ultimate criterion; while it has the additional misfortune to be a term which is used with varieties of meanings.[9] To clear up the case, it is quite necessary to start with a distinction between what can be represented by the sensitive imagination and what can be represented by the intellect strictly so called.

(*a*) As regards the sensitive imagination, what cannot be pictured by it need not, on that account, be impossible or untrue; else all our highest truths would straightway be undone. Contrariwise, what can be roughly pictured by the imagination may, as a concrete fact, be quite incapable of realization. A chiliagon, the square of 123456789, a mathematical straight line, the morality of an act, are all

---

[9] *Hamilton's Reid, Intellectual Powers,* Essay iv. c. iii. p. 377; Mill, *Examination,* c. vi.; *Logic,* Bk. III. c. vii.; Spencer, *Psychology* Part VII. c. xi.; Balfour, *Philosophic Doubt,* c. x.

objects with which the intellect may most accurately deal; but they all baffle accurate imagination by the sensitive faculty. On the other side, in a rough way, the imagination can form a sort of outline picture of a man standing on a single hair of his head, of Atlas supporting the world, of the cow jumping over the moon,—all which feats the intelligence pronounces physically impossible. They ought not, therefore, to be called without qualification, as they sometimes are, conceivable; for the conception never traces out the whole details, or it would find itself brought across absurdities. It follows that the possibility or the impossibility of picturing the opposite will not serve as the last, universal criterion of truth,—a conclusion for which we have already found sufficient reason, when we were considering the criterion afforded by verification through the senses.

Nevertheless, just as verification through the senses, in its own order, is an excellent and practically indispensable test of scientific theory, yet never so that mere sensation is the ultimate criterion of intellectual truth; in like manner all that Mr. Tyndall has eloquently uttered about the scientific use of the imagination in visualizing the minute processes of nature, must be granted to the full measure of the truth contained in his declarations. But sensitive imagination is not the last test of certainty—of the universal proposition in its universality, of the spiritual truth in its spirituality, nay, not even of the sensitive fact as stated in strict propositional form. A highly important consequence is the revelation of the truth, that with many persons their so-called intellectual difficulties against the Trinity, the existence of the soul, and the life after death, are not really intellectual difficulties at all, but difficulties of the imagination in its vain effort to picture the unpicturable. The proof is, that such people have no arguments to plead; only a baffled imagination.[10]

(*b*) The question must now be narrowed down to intellectual inconceivability; in which shape it calls for yet another distinction. If inconceivability of the opposite is taken negatively, for a mere impotence, it is not the ultimate criterion; for obviously the mere inability of a finite mind to see how a thing could be otherwise than as conceived by it, is no proof that the thing could not be otherwise. The simple incompetence of the spectators to conceive how a conjurer can do otherwise than betray certain indications, in some piece of sleight of hand, does not prove that he cannot avoid the betrayal. The point is too clear to allow of serious dispute, unless a man has the self-assurance to fancy, that

[10] Hume is of some use here: "A future state is so far removed from our comprehension, and we have so obscure an idea of the manner in which we shall exist after the dissolution of the body, that all the reasons we can invent, however strong in themselves, and however much assisted by education, are never able with slow imaginations to surmount this difficulty, or to bestow a sufficient authority and force on the idea. . . . *Except those few, who upon cool reflexion on the importance of the subject, have taken care by repeated meditation to imprint upon their minds the arguments for a future state,* there scarcely are any who believe the immortality of the soul with a true and established judgment," say rather, with a conviction which they can defend in set terms. (*Treatise*, Part III. sec. ix.)

there is no possibility beyond his powers of conception. We are left, therefore, to deal with positive inconceivability. What for positive reasons is seen to be such that its contradictory is impossible, implies more than a mere *impotence to conceive:* it implies a *power to perceive* that something cannot be. That must be true, the opposite of which is thus seen to be inconceivable. But it is a clumsy choice to pick out precisely the inconceivability as the ultimate criterion; for the more important element is the positive conceivability, or the evidence that something is as we see it to be. Whoever judges that something certainly is, implicitly judges that under the circumstances the opposite is inconceivable; the thing must be so and cannot be otherwise, however contingent may have been the fact of its realization. Here, however, what best deserves to be called the criterion of the judgment is its objective evidence. It is not primarily; because we cannot conceive the opposite, that we believe that two and two make four; but because we perceive the necessary identity between twice two and four. Even when a proposition is said to be proved negatively, the case is the same. In the *reductio ad absurdum*, and in the proof by exclusion of all hypotheses but one, positive conceivability is still the guide; evidence is the criterion.[11]

Inasmuch, then, as Mr. Spencer's criterion agrees with the one to be advocated in the next chapter, there is nothing to dispute with him; inasmuch as

[11] Mr. Bosanquet argues elaborately for a certain priority of the affirmative over the negative judgment (*Logic*, pp. 294—297).

it is vague, inadequate, and incorrect, it is to be repudiated. Besides those already indicated, one great flaw in it is its admitted fallibility, on account of which the author affirms that the less frequently his "universal postulate" enters into an argument, the better, for the less is the liability to error. Every use of the criterion is a fresh possibility of mistake. This premised, his rule is: Reduce any proposition to its simplest statements; then apply to each the test of the inconceivability of the opposite: the result is the nearest approach you can make to truth, while your dangers of having gone wrong are to be estimated by the number of times you have had to use your criterion.[12]

8. Here must end the review of criteria to be rejected; and from what has been seen, one conclusion impressed upon us should be, that the real criterion will have to accord with what we know to be the real nature of human intelligence. If a man steadily refuses to rise above the standard of associated sensations and their residues, if he will not ascend beyond the conception of *L'Homme Machine*, he can never hope to find a test of genuine certitude, for he is tied down to mere empiricism, or the doctrine which builds up knowledge out of mere associated ideas of experience, without any substantial soul that has an active power of intelligence. In a good sense we are all empiricists. The school-

---

[12] Compare with this theory Hume's view, that in strict reasoning every successive revision, by the mind, of its own fallible judgment, ought to reduce the mere probability with which it started to less and less dimensions, till nothing is left. (*Treatise,* Part IV. sec. i.)

men admit no innate ideas, no knowledge which has not an origin in experience; yet they are not what we call pure empiricists. They strongly maintain that the Leibnitzian *salvo* to a famous empirical rule is not mere verbiage, but expresses an important fact. As every one knows, to "Nought is in the mind which was not previously in the senses,"[13] Leibnitz added, "Save the mind itself,"[14] a most substantial addition against those who speak as though mind were a mere series of phenomenal states inherent in no substance. What seems a truism becomes really an important truth in opposition to those who deny it either formally or equivalently. The schoolmen make much of the doctrine that the intellect is no "mere abstraction turned into an entity," is not a mere name for the aggregate of all our ideas, but a principle of action, present from earliest infancy, though not ready to come into proper play till certain material conditions have been developed. In its activity, however, human intellect is subject to a condition analogous to that *inertia*, whereby matter does not act unless acted upon. Mind cannot act without some initiation on the part of the senses. Many points are left obscure, but what we gather with certainty from the interpretation of experience is, that the same soul which shares in eliciting the sensation, on the occurrence of the sensation frequently proceeds to a corresponding act of intelligence; and that intelligence, once possessed of ideas, has a large fund of

[13] "Nihil est in intellectu quod prius non fuerit in sensibus."
[14] "Nisi intellectus ipse."

power peculiar to itself, whereby it is enabled to push its knowledge far beyond the bare sensitive data. No doubt these data always form some limit to intellect, in such sort that the physicist must be perpetually feeding his mind with new observations; but on this account to deny the special power of intellect to enlarge upon its original data, is simply preposterous.

Consider the case of a man who has been a great observer, but not much of a thinker: if suddenly he becomes blind, and spends the rest of a long life in elaborating his acquired materials, what vast progress he may make in real science! Consider, again, the ample and objectively valid results due to geometry, synthetic and analytic; to mathematics generally; to mental and moral philosophy; and it will appear how mighty is that action of thought which supervenes upon sensation, and carries its conquests into regions not less real because their objects are not able to act on the sense organs. As the acute disciple may pass in thought beyond what his duller teacher tells him, so and still more may intellect pass beyond its source in sensation. It is, therefore, the veriest perversity to limit reality to the data of sense, and to declare all besides to be mere "symbolism," of no value except so far as it can be reduced back again to its sensible beginnings. Intellect is always valid so long as it proceeds in the only way which is intelligent, namely, not by blind mechanism or instinct alone, but with insight, seeing its way as it goes. Viewing it thus, we shall reach a criterion of certitude.

But for pure empiricists, with all their boasted adherence to the most literal realities, there is nothing left but that blank result, which Mr. Huxley says cannot be disproved—an empty idealism with no assured basis of reality. Their "objective and subjective sides," their "phenomena of the *ego* and phenomena of the *non-ego*," their "faint and vivid aggregates," all turn out to be mere shadows—shadows of the Unknowable, that is, of the Unthinkable, that is, of Nothing. Brahm, or Buthos, or Chaos, or the Mundane Egg, were names accounting for the universe of which we are conscious just as validly as do some recent speculations, which are supposed by their authors to be far above the old mythologies. In face of such disastrous philosophizing, we may well be moved to search after some really valid criterion of truth.

### Addenda.

(1) It would be small satisfaction to be told, that the laws of our nature are such as to compel us to accept certain propositions, if meanwhile our enforced belief could not be shown to rest on any rational grounds. Falstaff would "give no man reasons on compulsion;" and the mind equally objects to take compulsions for reasons. Are the Scottish school guilty of attempting this violence? Reid is not unfrequently accused of basing science on common sense, and common sense on blind instinct; but it is far from correct to say that this is his doctrine throughout his works. Many passages undoubtedly there are, which naturally enough lead to the unfavourable interpretation, and which, if they were not counterbalanced and even re-

tracted by opposite declarations, would deservedly bring his system under absolute condemnation. Neglecting what cannot be approved, let us, at present, show Reid on his commendable side; in places, at any rate, he asserts, not simply *necessity*, but *mental necessity*, which latter is a very different thing from *blind necessity*.

In the chapter on *Common Sense*,[1] passages like the following are found to redeem the author's reputation: "The same degree of understanding, which makes a man capable of acting with common prudence in the conduct of life, makes him capable of discovering what is true and what is false *in matters that are self-evident, and which he distinctly apprehends*." This contrasts strongly for the better with Hume's doctrine, that our faculties suffice for guidance in practical life, but not for the acquisition of rational truth. Reid continues: "All knowledge and all science must be built *upon principles that are self-evident*, and of such principles every man who has common sense is a competent judge, when he conceives them distinctly. We ascribe to reason two offices or two degrees: the first is to judge of things self-evident, the second to draw conclusions about things that are not self-evident from those that are. The first of these is the province, and the sole province, of common sense." And in the opening chapter of the Second Essay he had said: "Evidence is the ground of judgment, and when we see evidence it is impossible not to judge."

To declare, therefore, without large qualification, that Reid ultimately makes intelligence an unintelligent impulse to believe, is an unguarded criticism, which has been written too exclusively on the strength of some passages that we must now consider.

[1] *Intellectual Powers*, Essay vi. c. ii.

The grounds for misconceiving Reid are not hard to find; a specimen of them may be seen in Essay ii. ch. xx. What he there calls dark and inscrutable is, not the act of belief itself, but the nature of this act—how it is that we have faculties at all, and that they can do such a wonderful thing as is involved in knowing? Blind belief, and blindness to the mode of working in the faculties—these are two vastly different things: the latter of which, not the former, is what Reid really wants to assert. The process, so far as conscious, is intelligent: its nature considered as something, in the broad sense of the word, *physical*, is beyond the grasp of consciousness.

But, unfortunately, Reid has gone too far in setting forth the mystery of knowledge, thereby giving to his adversaries some foundation for the worst charges they bring against his doctrine. For instead of regarding the process as one competent to nature, he signifies that sensation, and its consequent idea, may have no more connexion than the will of the Creator that one should follow the other in definite order. "Whether they are connected by any necessary tie, or only conjoined in our constitution by the will of Heaven, we know not." No doubt this suggestion of occasionalism, or of the doctrine that definite conjunctions of created objects are merely the *occasions* upon which God acts on them in definite ways, is to be regretted; for it shows a readiness to take knowledge out of the sphere of natural causation, whereas we have good reason to regard it as a natural product. Reid, however, does not allow that his teaching thus removes knowledge from the domain of nature, but herein he is hardly consistent. We cannot more favourably take our leave of him, than when he is speaking so thoroughly in accord with our own doctrine as are these words of his: "That there are just grounds for belief may be doubted by no man who is not a sceptic. We give the name of evidence to whatever is a ground of belief. To believe without evidence is a weakness which every man is concerned to avoid. Nor is it in a man's power to believe anything, longer than he thinks he has evidence. What this evidence is, is more easily felt than described. It is the business of the logician to explain its nature, but any man of understanding can judge of it, and commonly judges of it right, when the evidence is fairly laid before him, and his mind is free from prejudice." [2]

Another representative of the Scotch school, Brown,[3] has expressions which some might seize upon to justify the common accusation that belief is made matter of blind impulse. "All belief," he says, "must alternately be traced to some primary proposition, which we admit for the evidence contained in itself, or to speak more accurately, from the mere impossibility of our disbelieving it, because the admission is a necessary part of our intellectual constitution." What is here called "speaking more accurately" is at least speaking more ambiguously, and is open to a construction which would make the doctrine condemnable. Perhaps the error is redeemed by referring the necessity to our "*intellectual* constitution:" for if the necessity is truly intellectual, it is not blind, but the effect of compelling evidence. Still Brown's case is rendered all the more suspicious because he denies the principle of efficient causality; and asserts, for such causality as he does admit, grounds which by his use of the word "intuition," and by his reference of this "intuition" to the bounty of the Creator, are rendered very insecure.[4] "We believe," he writes, "in the uniformity of nature,

---

[2] *Intellectual Powers*, Essay ii. c. xx.   [3] *Human Mind*, Lect. xiii. xiv.
[4] *Inquiry into the Relation of Cause and Effect*, Part I. sec. ii.

not because we can demonstrate it to others or to ourselves, but because it is impossible for us to disbelieve it. The belief is in every instance an intuition, and intuition does not stand in need of argument." Undoubtedly real intuition is immediate, not reached through the medium of argument; but Brown's view of intuition is peculiar.

If Brown is unsatisfactory, so too is Hamilton.[5] He teaches that knowledge rests on insight, belief on feeling; that the one cannot exist without the other; and that any definite act takes one name or the other from the element which is predominant. But he puzzles us when he goes on to say: "What is given as an ultimate principle of knowledge is given as a fact, the existence of which we must admit, but the reason of whose existence we cannot know." So far we might interpret him benignantly; but the next sentence is hard to take in good part. "Such an admission, as it is not knowledge, must be a belief: and thus it is that all our knowledge is, in its root blind, a passive faith, in other words, a feeling." This apparent basing of the element of "insight" on the element of " blind feeling " is very misleading: and the difficulty is increased by all that Hamilton has written about a belief of ours, the object of which he regards as inconceivable, involving not a conception, but a negation or impotence of conception, *e.g.*, "the infinite is conceived only by thinking away every character by which the finite was conceived: we conceive it only as inconceivable."[6]

Those who wish to see some defence of this writer may consult Professor Veitch's *Hamilton*, and Mansel's *Philosophy of the Conditioned*. The latter offers, as a key to a large part of the position, the following

[5] *Logic*, Lect. xxvii. Note A on Reid, p. 760; Discussions, p. 86.
[6] *Logic*, Lecture vi.

suggestions: "To conceive a thing as possible, we must conceive the manner in which it is possible; but we may believe in the fact without being able to conceive the manner. Had Hamilton distinctly expressed this, he might have avoided some very groundless criticisms, with which he has been assailed, for maintaining a distinction between the provinces of conception and belief." This hardly accounts for such a notion as we have of the infinite being called a mere "impotence of thought," "the negation of a conception:" nor is that account fully rendered even when we have further taken into consideration Hamilton's doctrine, that to *conceive* is to *comprehend under a class*.

On the whole, the Scottish school cannot be acquitted of blame, yet are perhaps less blameworthy than some of its critics have supposed. What it is popularly taken to teach, but what is not exactly its doctrine, is the suicidal theory, that there is a practical common sense, which sets reason at defiance, and is rightly thus defiant. Pascal expresses the same opinion in his famous sentence: "Nature confounds the Pyrrhonists and reason the dogmatists. Our inability to prove a truth is such as no dogmatism can overcome; and we have an apprehension of the truth such as no Pyrrhonism can overcome."

(2) When it is said that not many philosophers in this country regard our knowledge as due to ideas communicated from above, it is to be remembered that the late Professor Green of Oxford, and some other kindred thinkers, depart from what we may call the natural tradition as founded by Locke, and approach nearer to Malebranche. As a specimen, take the theory of Professor Green,[7] which it is difficult to give very intelligibly; but a few hints will suffice. He describes

[7] See more on the subject, Bk. II. c. ii. Addenda (3).

our process of learning as a gradual realizing of "the universal mind" in the "finite mind." First there is "a spiritual activity," which produces nature as a system of knowable and known relations, which relations cannot exist except as objects of consciousness. Then, part of this universal system of relations, known to the Universal Consciousness, also becomes known to finite intelligences, which "are limited modes of the world-consciousness," in some non-pantheistic sense of the terms. "The source of the uniform relation between phenomena and the source of our knowledge of them, is one and the same. The question, how it is that the order of nature answers to our conception of it, is answered by the recognition of the fact, that our conceptions of the order of nature and the relations which form that order, have a common spiritual source."[8]

(3) In denying to consistency the rank of the ultimate criterion of certitude, we must not in any way detract from its real dignity. Rather we ought to do our best to assert its true rank, in these days when system and coherence are often despaired of, and the best we can do is supposed to be to lay hold of a few "vital ideas." It is a sign of the times that a prophet in America could coolly write to a prophet in England, as Emerson[9] to Carlyle, in strains so characteristic, and so little scandalizing to a large body of admirers: "Here I sit, and read and write with very little system, and as far as regards composition with the most fragmentary result, paragraphs incomprehensible, each sentence an infinitely repellent particle."[10] The same author records in his journal: "I hate preaching; it

---

[8] Green's view may be seen compendiously in his Introduction to Hume's Works, § 146 and § 152.

[9] *Ralph Waldo Emerson*: a Biographical Sketch. By A. Ireland, pp. 27, 30, 110, 111, 124—129.

[10] *Ib.* pp. 27, 30.

is a pledge, and I wish to say what I feel and think to-day, with a proviso that to-morrow, perhaps, I shall contradict it all."[11] Speaking apologetically, he says: " It strikes me as very odd, that good and wise men should think of raising me into an object of criticism. I have always been, from my very incapacity of methodical writing, a chartered libertine, free to worship and free to rail, lucky when I could make myself understood, but never esteemed near enough to the institutions and mind of society, to deserve the notice of the masters of literature and religion. I well know there is no scholar less willing and less able than myself to be a polemic. I could not give an account of myself if challenged. I could not possibly give you the arguments you so cruelly hint at, on which any doctrine of mine stands."[12] His method of composition answered to the rest of the man. His habit was to go out almost daily and hunt after a thought; then coming back to record the day's capture in a book. So day by day he added to his list of stray ideas. When the time came to deliver a lecture, he went to his thought-record, strung a lot together like beads on a thread, with little care for definite harmonious result. The picture of one who so little valued consistent wholes is worth holding up as a warning to the present generation, in which so many, despairing of the reduction of their ideas to unity, set little store by consistent, systematic thought. Provided a man is clever, bold, and outspoken, he may pass for a great thinker; as is the case with many a mischief-worker like Diderot, of whom De Lamennais testifies, *Il nie tout, croit tout, et doute de tout, au gré de son imagination ardente et mobile.*

It is notable that Emerson was one of the first to hail Walt Whitman as a great poet, no doubt for verses

[11] *Ib.* p. 110.   [12] *Ib.* pp. 124—129.

## ULTIMATE CRITERION OF CERTITUDE.

like these which are culled from various "poems" in Rosetti's collection for English readers :

> I make the poem of evil also, I commemorate that part also.
> I am just as much evil as good, and my nation is.
>       And I say there is in fact no evil ;
> Or if there is, I say it is just as important to you, to the law, or to me as anything else.
> And I will show there is no imperfection in the present, and can be none in the future.
> What will be, will be well—for what is, is well.
> The difference between sin and goodness is no delusion.
> Whither I walk I cannot define, but I know it is to good.
> The whole universe indicates that it is to good.
> To me there is just as much in ugliness as there is in beauty.
> Of criminals, to me any judge or any juror is equally criminal,—and any respectable person is also—and the President is also.

Some may say the context will explain all these utterances: but that is not a plain man's experience, who finds one of the most intelligible and truthful of the verses to be this:

> Now I perceive I have not understood anything—not a single object —and that no man can.

Unfortunately, there are those other declarations to be got over, that obscure the little bit that seemed so obvious :

> As for me (lorn, stormy, even as I, amid these vehement days),
> I have the idea of all, and am all, and believe in all :
> I adopt each theory, myth, God, and demigod :
> I believe materialism is true, and spiritualism is true—I reject no part.
> I see that the old accounts, Bibles, genealogies, are true without exception.
> I assert that all past days are what they should have been,
> And that they could nohow have been better than they were,
> And that to-day is what it should be.

One reason for insisting on the First Principles of Knowledge is to prevent men like Walt Whitman from becoming the poets either of the future or of the present.

# CHAPTER XIII.

### EVIDENCE AS THE ULTIMATE OBJECTIVE CRITERION OF TRUTH.

*Synopsis.*
  1. The nature of human knowledge, and the consequent nature of its objective criterion.
  2. We have to show that this is *evidence*. What we mean by evidence.
  3. Proof that evidence is the ultimate objective criterion.
  4. Confirmation of the proof from animal instinct.
  5. A series of objections, serving to bring out more clearly the meaning of evidence as a criterion. (*a*) The criterion of evidence means judgment by appearances. (*b*) The criterion is a tautological, "that is, certain which is evident;" whereas we want a rule to settle what in every case is evident—not a declaration that the evident, when found, is the true. (*c*) How can abstract truths, and truths about mere possibilities have an objective reality, when they exist only as terms of the mind?
  6. The complicated nature of evidence.

*Addenda.*

OFTEN because they have expected too much from a universal criterion of truth, philosophers have declared that no such thing is possible. While some affirm that there are innumerable criteria for different cases, but no common criterion for all, others have gone further and proclaimed absolute certainty to be beyond human attainment. The question is undoubtedly difficult; and yet difficulties will yield to a patient examination of what it is we experience when we have these states of certainty, which previous propositions have shown to be sometimes ours.

1. The subject of a criterion has so many ramifications, that we must pick out what part precisely of the problem is to occupy our attention. And first it will be well to quote the very words of some of the schoolmen, in which they describe the process of knowing, and therefore the process of acquiring certitude, as involving acts of *conception*.

The schoolmen, to show that knowledge is no mere subjective fact, insist upon its origin in us by way of a conception and birth, and of double parentage.[1] Knowledge is generated by subject and object together: "Whatever object we know, this in union with the cognitive faculty generates within us the knowledge of itself. For knowledge is equally the product of both. Hence when the mind is conscious of itself, it is the sole parent of its self-knowledge; being at once the knowing and the object known."[2] The union of object with subject must be brought about "either by means of its own essence or by a similarity between them."[3] Thus teaches St. Thomas. In the same sense is the teaching of

[1] Cf. Kleutgen, *Philosophie der Vorzeit*, I. § 22.
[2] "Omnis res, quamcumque cognoscimus, congenerat in nobis notitiam sui. Ab utroque enim notitia paritur, a cognoscente et cognito. Itaque mens, cum seipsam cognoscit, sola parens est notitiæ sui; et cognitum enim et cognitor ipsa est." (St. Thomas, *De Trinit.*, l. ix. c. xii)
[3] "Sive per essentiam suam sive per similitudinem." (*Idem, De Veritate*, q. viii. a. 6.)

Suarez: "The cognitive power is in a state of indetermination as regards the production of this or the other object: hence to be determined to a particular act of knowledge, it needs to be placed in a certain relation with the object."[4] In the same way Silvester Maurus argues, that knowledge must be the joint product of faculty and object: as a vital, assimilative act it must be the work of the intellect; but for its determination to one definite similitude rather than to another it must be dependent on the object.

This doctrine, that human knowledge results from faculty as determined by object would be simple enough, if the intellectual object could be shown always to work upon the intellect, as a luminous body upon the eye. But an appeal to examples shows that the case is otherwise. According to St. Thomas and the Thomists, it is truer to say that, the intellect illuminates its object than that the object illuminates the intellect; evidence does not simply pour in upon the mind from outer things, but the intellect has rather to furnish its own light of evidence. Hence Lepidi writes: "The criterion *whereby* the mind judges is the faculty of judging; the criterion *according to which* it judges, is the rule or norm of truth, in other words, that inner light whereby an object becomes evident."[5] He

[4] "Potentia cognoscitiva est indifferens ad operandum circa hoc vel illud objectum; et ideo, ut determinetur in particulari ad cognoscendum, indiget conjunctione aliqua ad ipsum objectum." (*De Anima*, l. iii. c. i.)

[5] "Criterium *per quod* intellectus judicat est ipsa facultas judicandi: criterium *secundum quod*, est ipsa regula vel norma veri, nempe lux illa interior, secundum quam res manifestatur." (*Logica*, p. 236.)

further adds: "This light has, so to speak, two aspects, one, in so far as it is in the soul which it informs and perfects; the other, inasmuch as it actually represents the object outside the mind."[6] The first aspect he calls *subjective*, the second *objective*: but what may disappoint the reader is, that this objective aspect seems really part of the subjective light, not an influence, an irradiation, a determination coming from the object. If only thought could be described as the direct reaction of the faculty under a directly intelligible impression from the object, it would be satisfactory: whereas, besides its own intrinsic difficulties, the scholastic account of how material bodies are brought to bear on the determination of thought about themselves, seems to deny all real action of such bodies on the mind. The problem is confessedly difficult,[7] and has been assigned, not to the logical, but to the psychological division of treatises in the scholastic system.

Having stated where the deeper difficulty lies, we may proceed to do enough for the establishment of an objective criterion of truth within the limits of our own treatise.[8]

[6] "Habet hæc lux, ut ita dicam duas facies, unam quatenus est in anima, quam informat et perficit; alteram quatenus rem extra animam actu repræsentat ac refert." (*Logica*, p. 361.)

[7] Kleutgen, ut supra.

[8] What the need of this criterion is, will the more manifestly appear, if we look into the writings of some of those authors, who not being downright Kantians, are considerably under the influence of Kant's doctrine that we must inquire rather how objects conform themselves to mind, than how mind conforms itself to objects, and that there are *a priori* forms of mind, such as substance and accident, causality and dependence, which, for aught we can know, may

The criterion, *quo fit judicium*, is clearly the intellect itself, and this we suppose given: but the objective criterion, *secundum quod fit judicium*, this in its ultimate and universal nature is what we have to investigate.

Now we shall avoid the difficulties above signalized, if we take the problem up at a stage to which all must admit that it advances, however they may dispute as to the means of this advance. All certitudes concern propositions, and, in last resort, propositions are to be decided, not by inference from others, but on their own merits. Our inquiry into an ultimate objective criterion may take this shape: What, in last analysis, is the objective character of all those propositions, which, when they come before the mind for judgment, claim from it, for their own sake, a firm assent? This character will be the *criterium secundum quod* of which we are in search.

2. It may be declared at once that *evidence* is the *objective character, quality, or property which we seek*: but since the manner of this is not obvious at once, we must have the courage to plunge into details.

---

have no validity except as conditions of our thought. Such a doctrine is ruinous to objective knowledge and is too much favoured by Mansel (*Prolegomena Logica*, c. iii. p. 77), who tells us, that "the laws to which our faculties are subjected, though perhaps not absolutely binding on things in themselves, are binding upon our mode of contemplating them." When we hear such language we are prompted to seek an objective criterion, which at the same time shall be consistent with the subjective law, *cognitum est in cognoscente ad modum cognoscentis*.

*Evidentia* is the Latin word used by Cicero[9] for ἐνάργεια, the root of which is found also in *argentum, argumentum*, &c.[10] The radical meaning therefore is to make clear, bright, distinct, conspicuous. Everything, actual or possible, as is proved in General Metaphysics, has its truth—its ontological truth; and the manifestation, or shining forth of this, is called evidence. Hence the speculation as to whether there are, perhaps, things-in-themselves, which have no relation whatever to any intelligence, is philosophically absurd. Ontological truth is co-extensive with all being, and whatever makes this truth apparent to the mind gives its evidence. Not all things are evident to us, or our ignorance would not be what it is: still several things do become to us immediately or mediately evident; and when we speak thus, we are using the word evident not in its popular use for what is easily perceptible, but in its technical use for what is perceptible, whether by easy or by difficult means.

Evidence, therefore, is that character, or quality, about proposed truths or propositions, whereby they make themselves accepted by the intellect, or win assent; while the intellect is made conscious, that such assents are not mere subjective phenomena of its own, but concern facts and principles, which have a validity independent of its perception of them. In saying, then, that evidence is the ultimate criterion, we are implying, that the criterion

---

[9] *Academ.* Lib. II. c. vi. n. 17. (Nobbe's Edition.)
[10] "Nihil clarius ἐναργείᾳ, ut Græci perspicuitatem, aut evidentiam nos, si placet, nominemus."

is not, as some have vainly imagined, an all-containing proposition, from which any other truth may be evolved; further, that it is not a proposition at all, but a character of all propositions which so come before the mind, as rightly and for their own sake to demand its assent. When the nature of this character has been discovered, of course it may be declared in a proposition, or enunciated as a principle, "Evidence is the criterion of truth." But the criterion in itself is not a proposition or principle: it is a quality found in all propositions or principles which we can rationally accept, for their own sake, and is the reason of that acceptance.

3. To prove now that there is an objective evidence, which experience tells us to be our ultimate criterion. It is taught in theology that God is the substantial truth and always knows all truth. He does not gradually arrive at His knowledge by the use of faculties determined in their activities by outer agents; eternally and immutably He has all knowledge, without increase or diminution.

But we are beings that start with no knowledge, and gradually acquire our stock by passing *de potentia in actum*, from potency to act. Moreover, this transition is not effected by mere internal evolution; the faculties must be roused and determined by something other than themselves. Each faculty has its own proper excitant to which alone it is responsive. The ear responds only to one generic mode of outer vibration, the eye only to another, the palate only to what seems to be a definite kind of chemical process, and so on with regard to the other senses. Our finite intellect, in like manner, responds only to some appropriate character on the side of the objects presented to it, whatever be the way in which that presentation is effected. Now this character is what we call objective evidence, because the term accurately describes the state of things revealed by the careful consideration of our own experience. Surely it is right to frame our theory on the analysis of experience: and what it teaches is, that we do not make truth, but take it, when it urges itself upon us in a certain way, such that we feel it to be something independent of us, existing before us, and giving the law imperiously to our course of thought. Consider the proposition : " Nothing can arise by chance, everything must have a sufficient reason." In viewing the terms here, we feel that the relation between them forces itself upon us by way of objective evidence: we as distinctly feel the pressure put upon intelligence by some reality other than itself, as we feel on our bodily organs the pressure of an external weight.[11] Of course we may

[11] This is the idea which Locke, with no great success, tries to bring out in answer to his own question, how do men know that their ideas really represent the conditions of things ? " Simple ideas," he replies, "since the mind *can by no means make them to itself*, must necessarily be *the product of things operating on the mind* in a natural way, and producing therein those perceptions, which by the wisdom of our Maker they are ordained to." (*Human Understanding*, Bk. IV. c. iv. § 4.) He adds that simple ideas "carry with them all the conformity which is intended, or which our state requires, for they represent to us things under *those appearances which they are filled to produce in us*." Words like these last convey to many readers the impression that Locke regarded knowledge too

view the case on the subjective side, and say that it is insight which carries us along.  True, but insight must have its object, and must feel the influence of that object.  Mere subjectivism would never so distinctly objectivize itself, never tell us so plainly that the truth we contemplate is valid for all intelligence, and that to no intelligence can it really be manifest, as a truth for it, that events may happen without an adequately efficient cause.  Objective evidence must here lend its aid.

The argument will not avail unless we recall the doctrines already laid down about necessary truth, and about the first condition of philosophizing, which is our assumed ability to reach objective truth.  But with these doctrines in mind, we shall be forced to admit the fairness of the analysis, which, from an experienced act of certitude, disengages objective evidence as the element forming

much after the manner of the passive reception of a stamp impressed on the faculties by outer agents; and he is certainly unsatisfactory in what he teaches elsewhere in the same book. (c. ii. § 14.) Here he asserts our knowledge of the outer physical universe to be beyond "bare probability," yet not equal to "intuition" and "demonstration." If he meant no more than that physical certitude is of a lower order than metaphysical, he would have been right enough: but he seems to allow the possibility that the former may not be a full certitude: "There can be nothing more certain than that the *idea* we receive from an external object *is in our minds:* this is intuitive knowledge.  But whether there be anything more than barely an idea in our minds, whether we can thence *certainly infer the existence of anything without us which corresponds to that idea,* is that whereof some men think that there may be a question made: because men may have such ideas in their minds when no such thing exists, no such object affects their senses."

the criterion. Those who deny such an element, or who deny to it its right position, will be found denying necessary truth and violating the first condition of philosophy, as also asserting principles which lead directly to universal scepticism. Thus they violate the implied agreement of all intelligent discussion, that whoever in the course of it enunciates principles which are the subversion of all rational disputation, should be thereby declared to have sufficiently refuted himself, and to be silenced for the future.

4. The proof that objective evidence is man's criterion of truth gains some confirmation from a contrast with animal intelligence. It is the commonly admitted opinion, that, whatever may be the process of animal instinct, it is not one of calculated means and ends. If the bee does build what is mechanically the best sort of cell, it is not because of perceived mathematical relations, nor because of the perceived fitness. Thus the process, by its contrast with our way of deliberately adapting means to ends, serves to bring out more clearly our mode of thought, and to emphasize the criterion of objective evidence.

5. The meaning of evidence as a criterion will be brought out into still greater clearness, if we run through a series of objections against the term and its use.

(a) First, it may be said to sanction a habit of judging by mere appearances, on the maxim, "That is evident which to me appears to be,"[12] yet the sounder

[12] "Evidens est quod videtur."

maxim is, "Trust not appearances." In answer, we reply that appearances always are what, under the circumstances, they ought to be, if we except moral deception on the part of a free agent; so that it is not the appearances which are false, but our erroneous interpretation of them. In a sound sense we may give the advice, "Judge by appearances," for they are all you have got to judge by; and they are always the manifestation of some truth, with the exception just mentioned. By evidence, however, we do not mean sensible manifestation alone.

(*b*) From a charge of deceptiveness we pass to a charge of futility or tautology. "Where is the use," says an opponent, "of settling that the evident must be accepted as true? Of course it must; but the criterion we want is one which shall tell us, in all cases, what is evident." We answer that such a criterion cannot be found, or logic would be the sole science pointing out in every instance where truth lies. The logical criterion, which takes the form of the highest generality, cannot discharge this office of omniscience. Yet the function it does discharge is useful. When logic says, Objective evidence is the criterion of truth, it does not leave the words unexplained: else they might convey to the hearer no more than a truism: but it makes them the outcome of an analysis of the act of certitude; and thus they receive a fulness of meaning, which redeems them from tautology.

(*c*) "Be it so," rejoins our opponent; "but at any rate that is wholly subjective which is wholly in the mind; now truths about mere possibilities are wholly in the mind, and all abstract, universal truths formally exist only as terms of the mind. They are truths in the mind, but where is the objective evidence, or outer reality to which mind conforms?" The only reply to the first part of this difficulty is got by borrowing the results of a distinct section in General Metaphysics; in which it is proved, that possibilities are not mere nothings, nor mere mental terms, but have a real foundation at least in the nature of the Supreme Being, and often more proximately in some actually created nature. Each of them has an *ens essentiæ*, though not an *ens existentiæ*. As to the second part of the proposed difficulty, the reality attributable to abstract or universalised truths will be proved later. That there is some reality in possibilities and generalized science every one must feel, however much he may be unable distinctly to formulate to himself wherein it consists. Still the mere unformulated persuasion ought to induce the pure empiricist to distrust his position, which will not allow him to regard science as real in the laws which it lays down.

6. A further difficulty stands over in the fact, that what we speak of under the one simple name of evidence, enters into concrete cases after a very complicated way, and is far from being one simple thing. We must distinguish different evidences. Evidence is sometimes immediate, and then it presents no difficulty: but sometimes it is mediate, and the steps of inference may be many and intricate. Both mediate and immediate evidence may be intrinsic to the case considered, as in the most

abstruse mathematical theorem: but sometimes the evidence is extrinsic to the truth acquiesced in, as in the case where an ignorant man accepts a scientific conclusion, not from any insight into how it was derived, but from the evidence he has of the trustworthiness of his informant.

Again, the way in which what we call "the evidence" for a case is made up of several evidences in detail, some of which tend in opposite directions, is instructive as to the meaning of the term. Suppose a man charged with murder; the items for the defence being (*a*) that the prisoner had no discoverable motive for the crime; (*b*) that his previous conduct gave no serious indication of a character likely to be guilty of excessive violence: (*c*) that there exists another man likely enough *a priori* to have committed the crime, but quite free from any demonstrable connexion with it: and the items for the prosecution being, (*a*) that the prisoner, and only he, can be shown to have been near the spot about the time of the murder: (*b*) that there was a blood stain on his clothes: (*c*) that the weapon used was a dagger, and he possessed a weapon of that kind, which he says he parted with months ago.

Here let us speak of the evidences, rather than the evidence. First, they consist of the arguments which fully prove, as we will suppose, the respective three statements, *pro* and *con*: thus we have six separate certitudes. The difficulty begins when out of these we try to derive a seventh, namely, the guilt or the innocence of the man. At once we get into the region of probabilities, the very character of which is that full evidence is wanting, and we are left to conjecture beyond the reach of proof. It is precisely the probabilities which point to contradictory conclusions: the evidences, strictly so-called, cannot conflict, for so far as there is evidence there is truth, and no truth can gainsay another truth. There is some way of reconciling all apparent conflict, though we may not be able to find it out. Advertence to complications like these, while it clears up our ideas about the practical use of evidence, takes away all misgiving from the circumstance, that in spite of our having an infallible criterion, we are yet fallible judges, who blunder oftentimes. Evidence is safe where it is sufficiently abundant and direct to the point: but evidence, scarce and indirect, may very well prove a fallacious means when employed by creatures such as we are. But of this in the next chapter. Here it only remains to add, in conclusion, that unsolved difficulties do not destroy a certitude once fully established; for probabilities disappear before a contrary certainty, no matter how preponderant their weight may have been as probabilities. If the highest probability were beyond all fear of a failure, it would be certainty, and not probability.

ADDENDA.

(1) Some schoolmen, besides the wider sense of evidence, use a narrower sense, according to which that only is evident, which has necessitating evidence, making the truth so clear that the mind cannot well

refuse assent. Such evidence does not exist in some instances, where an element of good will is requisite for arriving at the right conclusion. In this sense we hear of propositions being certain, but not evident.

(2) The schoolmen describe material objects as being in themselves not immediately intelligible: hence they deny that a material object can efficiently act on the mind; and many carry this denial even as far as to include under it mediate action through the sense-image in the brain. Hence a long discussion about the *illuminatio phantasmatis* and the production of a *species intelligibilis*. The matter must be left to psychology; but it so closely bears on the thesis about objective evidence, that to fail of noticing the near connexion would hardly be right. At any rate we can always insist that intellect, be its object material or not, is guided by objective law, not by mere subjective evolution, independent of an object; and that the senses have a demonstrative influence on the objective side. We need not, therefore, call in any mystical theory, such as that apparently suggested in Mr. Wylde's *Physics and Philosophy of the Senses*, where we read, that "the whole of our intercourse with nature is literally the connexion of mind with mind, between the Great Mind and the mind of His creatures, not by miraculous means, but by and through the operation of those ordinary laws, of which He is the present and sustaining principle." If this means only that God sustains and cooperates with all secondary agencies, it is correct; but if it implies that secondary agencies are not adequately operative in their own manifestation, it is erroneous.

(3) The criterion is laid down for our ordinary knowledge, not for any supernatural or preternatural communications. Neither does it concern those things which must, in part at least, be matters of personal

taste, without an absolute objective standard, such as the choice between two recognized styles of architecture, of music, or of painting. Preferences in these matters must be largely referable to subjective conditions; and the extravagance is, when a man insists on making his own private likings a law for others, who are just as competent to decide for themselves. The misery is, that so many people, especially in matters of variable taste, are so insistent upon an invariable conformity to their favourite standard, which has no valid claim to be exclusive. Because the matters are so little to be fixed by argument, therefore strength of assertion is called in to supply for proof.

(4) A curious phenomenon of imagination or emotion which some seem to mistake for a failure of intelligence, is exhibited in cases where men, out of fear, will not act when reason clearly tells them it is safe to act. Thus some will go to great trouble rather than step over a serpent, which they know to be dead; others cannot be persuaded to take an eel off a fish-hook, on account of its likeness to a serpent; and others will not go near a corpse, which they are intellectually convinced will do them no harm. At least these examples do not diminish the rank of evidence as a criterion for assents of the mind, whatever they may do against man's character for reasonable conduct.

# CHAPTER XIV.

### THE ORIGIN OF ERROR IN THE UNDERSTANDING.

*Synopsis.*
1. Ignorance the root of error. How we begin in ignorance, slowly acquire some knowledge, but never cease to be in many ways ignorant.
2. The scholastic theory about error is, that the intellect is *per se* infallible, *per accidens* fallible: and that undue influence of the will is exerted in the case of error.
3. The scholastic theory taught outside scholasticism.
4. Supplementary considerations to complete the theory. (*a*) Dependence of the intellect on organic conditions, which are liable to disturbance. (*b*) The force of habit on the interpretation of sensation by the intellect. (*c*) The piecemeal, defective way in which we obtain evidence.
5. The scholastic theory re-stated and modified by the supplementary remarks.

*Addenda.*

THE next problem pressing for solution is to settle how, in spite of the fact that in evidence we possess an unerring criterion, yet we do err: so that intellectually, perhaps as much as morally, *humanum est errare*. The difficulty weighs heavier upon us than it would on those who, with Grote, believe that " no infallible objective mark, no common measure, no canon of evidence recognised by all, has yet been found." We who assert such a canon, have to explain how intellectual error is not only possible, but of constant occurrence, being sometimes practically inevitable.

1. Ignorance is not itself error; but it lies at the root of error; inasmuch as, while an Omniscient Being cannot err because of His omniscience, a creature, because his knowledge is but partial, is exposed to the risk of forming false judgments. It is the little knowledge that is the dangerous thing.[1]

We must, then, advert to the fact of our ignorance—how we begin in blank ignorance, very slowly emerge from the universal darkness, and never reach the full blaze of knowledge complete. Our knowledge is always a small sphere of illumination enclosed in an infinite sphere of obscurity; and the more the former grows, the more does its wider contact with what is without make it sensible

[1] There is a certain semblance of truth in the caution given by Rousseau: " Remember, always remember, that ignorance has never done any harm, and that only error is mischievous; that a man is not led astray by what he does not know, but by what he wrongly fancies that he knows." (*Emile*, Lib. III. in initio.) In a later passage towards the end of the same Book III., he returns to the subject: he says that all our errors come from judging; if only we had no need to judge, we should avoid error, and should be happier in our ignorance than our knowledge can make us. He thinks that learned men have less of truth than the unlearned, because each truth that they take up is accompanied with a hundred false judgments; so that the most famous of our learned societies is only a school of falsehood, and there are more mistakes in the Academy of Sciences than among a body of Hurons. " Since then the more men know the more they fall into mistakes, the only way to escape error is ignorance. Never judge, and you will never deceive yourself. This is the lesson of Nature as well as of reason." He adds, however, that as circumstances force us to form judgments, we had better study how to form them rightly.

of its own limitations. Consider our personal history. For years the brain is not fit to serve the uses of higher intelligence: and when what is called the age of reason has arrived, long years of education are still needed to form the faculties into efficient working powers. Again, when at the age of about twenty the condition of pupilage is over, a young man is told, as a parting piece of advice, that he is not a learned Doctor, but that he has the outfit necessary for setting about the work of becoming learned; and that even in its fully developed state, human learning is an ornament which is to be worn with a modest appreciation of its perfection. Moreover, the knowledge which a man is said to have acquired is not always ready at need, as a schoolboy doing his Latin exercises will testify: and the knowledge that does not come up when wanted, is for the moment equivalently ignorance. Such is the extent of our ignorance.

2. Ignorance being supposed, the transition from it to error has to be studied: and in the course of our explanation we shall come across the promised account of how it is, that while judgment is defined as the full perception of the connexion between subject and predicate, yet judgments may be false.[2] It is the theory of the scholastics that intellect in man is *per se* infallible, *per accidens* fallible; or more accurately, *per se non fallitur, per accidens fallitur*. For it is *per se* fallible only inasmuch as, being *per se* a finite intelligence, it is of its own nature *exposed to the possibility* of going astray, but it does not simply of its own

[2] See Bk. I. c. iv. p. 52.

nature actually *go astray*. Similarly the finite will is *per se* peccable in so far as it is *exposed to the possibility* of sinning, not because *per se* it sins. The intellect, as such, is moved only by its own proper object, which is evidence; and as evidence is the unfailing criterion of truth, the action of the intellect, strictly so called, is never erroneous. Intellect acting *per se* goes only by insight, and insight is always right. Thus insight *per se* can no more assent to anything but truth, than the ear proper can be sensible to anything but sound. But intellect, so far as it is subject to the undue action of the will, may be moved to go beyond or against the evidence it has at its disposal. This theory will be defended as in substance correct, though it may be usefully supplemented with some further considerations, much urged by modern writers. First, however, it may gain for itself a little more attention, if it is shown not to be an exclusive property of scholasticism, but to be owned likewise by thinkers of various classes. A multiplicity of approvers may induce some not to pass over the theory in contempt.

3. Hamilton was fond of quoting the line from Manilius—whom we may take as our oldest witness, returning after a moment to Hamilton himself—*Nam neque decipitur ratio, nec decipit unquam.* Second in order we will take Descartes, who assuredly had no scruple in breaking loose from the scholastic bonds of his early educators, whenever it suited him. He holds firmly to the doctrine that error springs from the bad use of the will, not from intellect left to itself. In the first book of the *Principia* he

writes :[3] " That we fall into error comes from some defect in the employment of our powers, not in our nature as such, or in the use of our free will. Since then we are aware that all our errors may be traced to the will, it may seem wonderful that we should ever be deceived, for nobody wishes to be deceived." Then he adds acutely: " But the will to be deceived is quite other than the will to assent to something which happens to involve error. And though it be true that no one is willing to be deceived, there is hardly any one who does not will assent to what contains error, though he be not aware of it."[4]

Another Frenchman,[5] Cousin, writes : " Pure error is impossible, and quite unintelligible : for error makes its way into the mind only by means of the truth which it contains."

Passing next to those who write in the English language, we may begin with the already promised quotation of Hamilton's opinion.[6] He holds that what we really and positively think cannot be

---

[3] "Quod in errores incidimus defectus quidem est in nostra actione, sive in usu libertatis, sed non in nostra natura. . . . Jam vero cum sciamus errores omnes nostros a voluntate pendere, mirum videri potest, quod unquam fallimur, quia nemo est qui velit falli." (*Principia*, Part I. nn. 37, 39.)

[4] "Sed longe aliud est velle falli, quam velle assentiri iis in quibus contingit errorem reperiri. Et quamvis revera nullus sit qui expresse velit falli, vix tamen ullus est qui non sæpe velit iis assentiri in quibus error, ipso inscio, continetur." (l. c.)

[5] See his twenty-fourth lecture on the *History of Philosophy*, where he treats Locke's theory of error : " La pure erreur serait impossible, et elle serait inintelligible : comme l'erreur ne pénètre dans l'esprit d'un homme que par le côté de vérité qui est en elle."

[6] *Logic*, Vol. II. Lectures xxix., xxx.

erroneous, and that error is rather a want of intellectual action than an intellectual act. Mansel[7] concurs with his master, and holds that "illogical thinking is no thinking at all." Dr. M'Cosh[8] is another consentient witness: "I cannot keep from giving it as my decided conviction, that while ignorance may arise from the finite nature of our faculties, and from a limited means of knowledge, positive error does, in every case, proceed directly or indirectly from a corrupted (?) will, leading us to pronounce a hasty judgment without evidence, or to seek partial evidence on the side to which our inclinations lean. A thoroughly pure and consistent will would, in my opinion, preserve us from all mistake." Finally, one who is not writing on philosophy shall join his voice to those of philosophers: "Mere sophisms or imperfect reasonings," says Mr. Lecky, "have a very small place in the history of human error; the intervention of the will has always been the chief cause of delusion."

4. This view that the will is the cause of error, supported as it is by so many authors, may be supplemented by some considerations much urged by modern writers—considerations which are, however, really supplementary, not contradictory to the theory propounded.

(a) One source of delusion is in the derangement of the nervous apparatus; and the nature of this perturbing action will require some detailed account.

[7] *Prolegomena Logica*, p. 250. See Hobbes on Error, *Leviathan*, Part I. c. v.; and Hume, *Treatise*, Bk. I. Part IV. § 1.
[8] *Intuitions of the Mind*, Bk. II. c. ii. § 2; Bk. IV. c. ii. § 2.

It is no new fact that a lesion in the material organ may result in stopping thought; and that on account of altered cerebral conditions a man may be in any one of the countless gradations between sleep and wakefulness, or between sanity and insanity. And as sleep has its dreams and insanity its delusions, so in the intermediate stages just mentioned there may be intermediate degrees of deceptiveness due to an abnormal state of nerves. Some people labour under the frequent recurrence of visual or auditory illusions, which they can calmly correct by data supplied through the other senses. When the inflow of sensations from the ordinary channels is cut off, there are patients whose minds become quite deranged by their own subjective phantasies, and who are restored to composure only by being brought from darkness to light, and by having their several senses fed with their usual supplies. They need the steadying influx of impressions from the outer world to prevent the inner life from upsetting its own balance. An excitable man suddenly deprived of his hearing in a public thoroughfare, would often grow quite bewildered for want of his customary guidance from the ear; and still more would this be the case if the deprivation was effected, not merely through an external stopping of the ears, but through some inner disorder of the nerves. Thus in many ways a disturbance of the normal working of the nervous system has its result in a disturbance of the mind, and erroneous judgments not unfrequently follow.

From the most general statement of the fact we may now come down to a particular law, which may be enunciated thus: Whenever in the brain extraordinary causes which are internal excite those phenomena which ordinarily are excited by familiarly known objects, there is a tendency erroneously to judge those objects to be present, though in reality they are not. Sometimes it is the vehemence of an idea which excites the sensible image, and at once the object is as if bodily present: at other times the action is rather from below upwards, and the abnormally roused sense-images call up their corresponding ideas. Here we are safe in asserting that we have an undoubted occasion for an erroneous judgment, as for example when the vivid thought of a departed friend has brought up his image in the brain, and he is declared to have been seen. Some, though not all, ghost stories may be so explained.

(b) Again, there is a second special law of delusion through the senses, the law of *the accidental miscarriage of customary interpretation;* and it differs from the first in not implying any internal derangement of the nerves. Ordinarily, what we actually at any time perceive is the merest item, compared with all that is at once filled into the object by inference or association. We catch sight of a plume and we at once supply a hearse; we observe a wheel moving, and we supply the whole carriage and its occupants. An odour leads us to assert the presence of oranges or lemons; a sound the presence of an organist and his instrument; a touch a broken bone beneath its muscular covering. The practical necessities of life drive us to make these short cuts by the aid of incomplete inference; for if we stopped fully to verify everything, we could not get through one tithe of our business. As a rule, our customary inferences from few data are right, but every now and then they are wrong; and whoever cares to play us a practical joke may probably succeed in doing so, if under familiar appearances he will present to us an object not usually associated with them in our experience. In the examples given above, while we do not say that the unusually produced sensations or sense-images are errors, we must say that they may be occasions of error, and sometimes of error practically unavoidable.

This is the moderate statement of the case, and contrasts with the immoderate statement of M. Taine:[9] "The two principal processes employed by nature to produce what we call acts of cognition, are the creation of illusions within, and their rectification. It is a point of capital importance that external perception is a true hallucination. When sense-objects really impress us we have first the sensations, which an hallucinated person has without real objects. The external perception is an internal dream, which proves to be in harmony with outer things. We have, when awake, a series of hallucinations, which do not become developed. This hallucination, which seems a monstrosity, is the very fabric of our mental life. Nature declines to instruct us. In recollection a

---

[9] *De L'Intelligence*, Part II. Liv. I. c. i. pp. 411, seq.

present image is taken for a past sensation. Just as, in external perception, simple, internal phantasms are taken for external objects, so in memory we see simple present images taken for past sensations, but corresponding by a beautiful mechanism to the exterior presence of real sensations. The history of sleep and of madness gives us the key to the waking state." Mr. Sully[10] has some remarks of somewhat like tendency, when he is speaking of the region of hallucination as a border-land between reason and insanity, or rather as forming the extreme confines in which these two regions are, as it were, blended. He adds that "in perfect normal perception we find in the projection of our sensations of colour, sound, and the rest, into the environment or to the extremities of the organism, something which, from the point of view of physical science, easily wears the appearance of an ingredient of illusion." We may be pardoned this speedy recurrence once more to the subject of the force which habit has in misleading us,[11] for the matter is once more strongly urgent upon our attention, now that we are engaged explicitly in giving an account of the origin of errors in the human understanding. Those who refer all judgments to repeated associations of ideas, naturally make much more of this source of error than we can allow; but we can allow that it is a source.

(c) Lastly, the criterion of evidence often fails to secure us from error, because we get our evidence piecemeal, in insufficient amount, and

[10] *On Illusions* (International Series), pp. 60, seq. pp. 111, seq.
[11] The subject has been already discussed in c. vii. pp. 124, seq.

often with only indirect bearings. If the evidence of each case were one simple thing, we should run no risk; but, as was observed in the last chapter, we usually have to deal with a complicated mass of evidences in the plural.

5. Examining next how the scholastic theory can accommodate the three supplementary considerations, we note first that all three elements, at least indirectly, come under the control of will, to a large extent. By force of will we can often resist or correct abnormal conditions of the sensitive system, and by force of will we can aggravate these conditions. Again, will has a large share as well in forming our intellectual habits, as in checking them. Lastly, will has its influence in setting us carefully, cautiously, and restrainedly to judge from complicated evidences, or in urging us precipitately to force a conclusion.

While, however, these several conditions are controlled by will, they have distinct influences of their own; and this is the reason why the theory, that error is due to will, seems not complete, unless they too receive special mention as factors of the whole.

In what sense, therefore, from our larger survey of the position, can we admit Hamilton's dictum, that "No error can be really thought?"[12] Are we to say,

---

[12] The Hamiltonian school adhere for the most part to thi doctrine. Thus, besides Mansel, we find Professor Veitch saying: "There is only one way of thinking by the understanding, that is, the legitimate way. Any other is mere illusion, not a reality of thought at all." (*Institutes of Logic,* p. 7.)

that he who honestly mistakes his neighbour's hat for his own, does not really think it his own? Not so; but what we may assert is, that in his way of forming this judgment there were some steps taken in which thought was a blank. The man never really thought out all the steps to the conclusion—" This hat is my own." He thought out part and filled in the rest by force of habit, association, or rash inference. And the like may be affirmed of every case of error. A man has worked out a long mathematical problem : he assents to the conclusion, but not from clear insight of what is involved in it ; his assent is given in trust that his working out of a long process was right at each step. But some step or steps there must have been which he never represented in thought, and so " the error was not really thought."

Somewhat in the spirit of these last explanations, it has been said, that if the old astronomers had only stated the limits under which they were speaking, their statements would have been correct. They assumed that there was an absolute upside, opposed to an absolute downside : they assumed that men could not stand on the earth if it were placed upside down : from these premisses their inference was valid, that the earth could not be revolving. From the hypothesis of a stationary earth, they rightly inferred the motion of the sun. Thus they never fully thought out the real problem, but an ideal problem which was consistent with itself. Not thought, but something else, carried them over some parts of their argument when they applied it to the actual system : but if they had put in their limits, then their view would have been hypothetical and right. Instead of taking the absolute form, "The earth is fixed, the sun revolves round it," their astronomy would have taken a hypothetical shape, "If certain suppositions are true, then the earth is fixed and the sun revolves round it."

To put the whole of this part of our doctrine summarily: the error assented to is either a contradiction in terms, and then it is clear that it has never been strictly an object of thought, or it is an error in a contingent matter, and then the final result may in some sense be said to be an object of thought, but at least its actuality has never properly been thought out to the full. We may really think that X was intoxicated when he was not; but we have never followed out in thought all the evidences for the fact. At some point, not thought, but another power, has effected a part of the process.

In this way Hamilton's saying, which is in conformity with the scholastic theory of error, if not made to mean more than it necessarily implies, expresses a useful doctrine. It corresponds to that which Descartes probably meant when he said that, if he was only careful always to follow clear ideas, and nothing else, he could never go wrong. Unfortunately he did not describe properly the criterion of clear ideas; but we may add the explanation, that clear ideas must mean insight into objective truth. Insisting on this insight, we necessarily assign a very different account of the genesis of error from that which is assigned by those who treat only of the mechanism or chemistry of ideas; of associations and dissociations, of affinities and repulsions between mental atoms. Once more it is seen how a philosophical explanation is dependent on the radical nature of a system; and how the followers of Hume are in their whole point of view at variance with truth. A theory so erroneous as Hume's can never render the right account of error, though it may serve as an illustration of it to an expounder who goes on true principles. On these true principles we have laid down our theory, that ignorance is a condition, but never by itself alone; the efficient cause of error, and never identical with error: that the ignorant mind is necessarily *fallible*, but not with the same necessity *actually false:* that the man who labours under incomplete and obscure ideas is essentially *exposed to the danger* of judging wrong, but does not so essentially *judge wrong in fact;* that habits and associations incline us to assert more than is in the evidence before us; and finally that the will exerts its power to urge on acts of assent or dissent, which the mere intellect of itself would not have made, because these being untrue, are not fit objects to decide an intellectual movement. The grossest mistake must have some element of truth in it; and, "falsehood is dangerous only from its possessing a certain portion of mutilated truth." Thus evidence itself helps to elicit the erroneous judgment; but it is precisely because, besides evidence, there are other forces at work, that the total result is a failure.

## Addenda.

(1) In saying that our ignorance is infinite compared with our knowledge, we must be taken as referring to the *details* which in any concrete enunciation are left to be filled in: for, of course, under the *generalized* terms Being, Substance and Accidents, God and Nature, we include all things in our knowledge.

(2) When distinguishing will from intellect, we require no more than such a distinction as all admit who allow that to know a thing is not the same as to wish it. This leaves quite intact the question whether the several faculties of the soul have a real distinction *inter se*, and from the soul to which they belong. Some of our modern English writers assert that every mental act contains an element of *thought, feeling*, and *volition*, the three constituents of mental life. It may be true that the intellect never embraces truth, which the will does not somehow, at the same time, embrace, at least for its truth's sake, though under other respects the will dislikes the object intensely. Yet, on no account could we admit the Malebranchian theory, that the assent in a judgment is the act, not of the intellect, but of the will.

(3) A further question is whether the action of the will in error is always free. Suarez[1] speaks as though it were; but allows such a *minimum* of freedom sometimes as would save from moral guilt. In accordance with his teaching, we hold the existence of countless limitations upon that freedom, especially in what are called "first motions of the will," the *motus primo primi* of theologians. Very often, at any rate, our errors are proximately or remotely due to an abuse of freedom:

---

[1] *Metaphysics*, disp. ix. § 2,

but we may refrain from saying that they are so always.

(4) The importance of the power of will in determining judgment has, besides a high speculative, an equal high moral importance. It is an undoubted fact, that the erroneous judgments of many persons are most culpable. We have only to note what an abatement of assertions there is, as soon as an ordinary talker is brought to book, and as it were put on his oath, to infer how very rash are a great mass of human assents. It is said that many would sooner have their good will than their sound judgment called in question: they prefer to confess a culpable negligence rather than an inculpable mistake. But the two departments are connected; so that a man cannot constantly be guilty of great wilfulness in his judgments, without intrinsically damaging his very power to know the truth. In the interest of his intellectual faculty itself he must exercise a most vigilant use of his will, as a determinant of his assent.

# THE FIRST PRINCIPLES OF KNOWLEDGE.

## PART II.
### SPECIAL TREATMENT OF CERTITUDE.

### CHAPTER I.

#### SHORT INTRODUCTION.

*Synopsis.*
1. Transition from the general to the special treatment of the subject.
2. (*a*) Substance and (*b*) Efficient Causality at the basis of the treatment.
3. Enormous difference between the point of view taken by pure phenomenalism and that taken by the schoolmen.

1. A DESCRIPTION of certitude in general has now been given; and it might be supposed that next, each of the several faculties concerned in the production of certitude would be taken separately, and shown to be a valid instrument of knowledge. This would fairly stand as the special treatment of the subject. But it is convenient to leave alone the question as to how many faculties there are, and how to

divide them; for a more serviceable method suggests itself. If it be established successively, that our sensations, our ideas, our consciousness of self and its affections, our memory, and our belief in the testimony of others, are all, in their own nature, means for putting us in possession of certitude, whatever may be their liability to occasional, accidental error; then, without any list of faculties, enough will be done to satisfy any reasonable requirements on the part of those who ask a detailed justification of our claim to real knowledge. Here is our work in this Second Part.

2. Before proceeding to the task proposed, it is quite necessary to make explicit statement of some doctrines about substance and efficient casuality, doctrines lying at the very root of any theory of knowledge, yet doctrines which do not belong to this treatise, but to that on General Metaphysics. Here, however, a brief declaration is almost imperative, in this country where Hume has such an influence.

(*a*) The notion of substance, which scholasticism upholds, is not what the school of Hume is apt to fancy. By substance is not meant a mysterious entity which cannot be reached, and is hidden away under a shell of merely phenomenal realities—whatever these may be—like an Oriental monarch, awful in his utter unapproachability. Listen to what are the essential demands of the schoolmen, who hold a very different doctrine. Many of them, it is true, do suppose, betweeen the quantity and the qualities of an object on one side, and their subject of inherence on the other side, a distinction so real, that it is second only to the distinction between substance and substance. At the same time, they admit that such real distinction is not contained in the primary notion of substance; that it is a secondary point of investigation, quite open, on merely natural grounds, to strong controversy.

But the primary notion of substance, the incontrovertible notion, the universal notion applying even to God Himself, Who is without accidents—this they place in what they call *perseity*. Substance is what exists *per se;* and what to exist *per se* means is brought out by a contrast, the validity of which cannot be gainsaid. We leave alone these accidents of quantity and quality which are supposed by some to be more than modal, and the nature of which is matter of dispute. We keep to what is indisputable; thought, volition, motion, rotundity, these are in some sense realities, and yet none of them can exist *per se*, all must inhere in some subject, and are really distinct from that subject at least modally, or, inasmuch as they are modes, which may, or may not, affect a thing, while that thing remains substantially the same. But they are only modes: no one yet ever came across rotundity existing by itself; no one ever met a piece of motion unattached, without a thing of which it was the movement. Similarly a wandering thought or volition, in the sense of an entity which is nothing but a thought or a volition, an isolated phenomenon, is an absurdity.

To recur again to examples. A cannon-ball is now

at rest, and now endowed with a most terrific velocity: in the one instance a child may support it, in the other hardly the strongest target that man can make will resist the momentum undamaged. Therefore the velocity has some sort of a reality not wholly identified with the substance, as such, of the ball. Again, the mind may rouse itself to intense thought, or yield to comparative quiescence; the thought is some sort of a reality not wholly identified with the substance mind. There is then at least one class of accidents, the modal, which are real, and which present some real contrast to substance. These suffice to enforce the definition: "An accident is that which exists in another, as in a subject of inhesion;"[1] where the precise degree of real distinction involved by the *in alio* may be left without further niceities of discussion. Mill has a glimpse of the truth, soon to be lost amid erroneous ideas about the unknown *substratum*. In the third chapter of his *Logic* he says: "Destroy all white substances, and where would be absolute whiteness? Whiteness without any white thing is a contradiction in terms."

As illustration of a doctrine, the full proof of which is to be sought in General Metaphysics, the above account must suffice to justify the assertion, that the radical notion of substance is intelligible and real. After the manner described,[2] "substance is that which exists by itself, and does not inhere in

---

[1] "Accidens est id quod existit in alio tanquam subjecto inhæsionis."

[2] "Id quod per se stat, et non inhæret in alio tanquam subjecto inhæsionis."

something else as in a subject of inhesion."[3] Realities cannot be inherent one in another indefinitely, like the earth supported by a rock, and that rock by another, and this by a third, and so on unlimitedly; in the end there must be something which exists *per se*. Now *per se* might mean self-existent, uncreated, unproduced; but here it does not mean that: *a se* is the expression used to signify underived existence. God alone is *a se*, and therefore also He is *per se*. How *perseity* can be assigned to creatures without denying their continuous dependence on the Creator is a difficulty which is briefly met by saying, that unless some creatures were *per se*, all would inhere in God as accidents of the Divinity: as parts of His total reality. This would be pantheism.

Whence it further appears that the primary idea of substance is not permanence under varying accidents. God is substance, though having no accidents. He is immutable; created substance, though it were annihilated almost as soon as created, would have been for the moment real substance.

Mr. Bain, therefore, is utterly wrong in saying that substance has no meaning; and Mr. Huxley, who says that "whether mind or matter has a substance or not, we are incompetent to discuss." But Mr. Spencer has got hold of a partial truth, when he holds, that "the conception of a state of consciousness implies the conception of an existence which has the state; we are compelled to think

---

[3] See Lepidi's *Elementa Philosophiæ*, Vol. II. Lib. II. sect. ii. c. i. For Mill's admissions, see the present volume, Bk. I. c. xi. Addenda.

of a substance, mind, that is affected, before we think of its affections:" and that "it is rigorously impossible to conceive that our knowledge is of appearances only, without at the same time conceiving a reality, of which they are the appearances."[4] It is idle to pretend that the necessary recurrence to substance is a mere association of ideas, or a mere grammatical notion. Grammar, it is true, distinguishes substantive and adjective; but so manifestly is this not the philosophical distinction between substance and accidents, that many nouns substantive confessedly stand for accidents, as velocity, rotundity, volition. Also, it is true, Aristotle teaches that the concrete substance, the *prima substantia*, πρώτη οὐσία, can never be predicated of anything else as of its subject; but what is this against the reality and the knowableness of substance? In the notion of substance we have got hold of the undoubtedly real. We do not lay bare a great mystery, as many suppose we pretend to do; but we do affirm a clear truth, which is elementary in the human understanding, and without which the mind is lost in nihilism.

(*b*) Efficient causality, like substance, is supposed to be a chimera by the disciples of Hume. Again let us oppose our doctrine to theirs. We waive the question whether there are any substantial changes in nature: but at least there are real changes, and a vast multitude of

[4] How far, however, Mr. Spencer is from holding the true doctrine of substance, will appear on reading *Psychology*, Part II, . . "The Substance of Mind."

them. Forthwith we take our stand on plainest and surest of principles. Nothing begins to be without a sufficient reason: real events are perpetually beginning to be in this world, which we familiarly style "a world of change:" the sufficient reason, or part of the sufficient reason, for a real change is an efficient cause. There are then real efficient causes, and we know that there are. We do not know *how* efficient causality ultimately acts, but we know *that* it acts. We may be silent as to the difference or the identity between substance and its powers: but on the reality of the powers we may not be silent. They clamour for recognition. If anything is certain in this world, it is that mere uniform sequence, without any idea of power, is an inadequate account of a real succession of events. Mill, after the manner of his school, seems to be confounding the primary with the secondary question, the question as to the reality of power with the question of the reality of its distinction from its substance, when he says with an air of apparent triumph: "It is as easy to comprehend that the object should produce the sensation *directly*, as that it should produce the same sensation by the aid of *something else*, called the power of producing it." If the reader will admit *substance efficiently active*, without any question raised as to an intermediate reality between the substance and its activity, he will admit enough for the purposes of the following discussions on the details of certitude. But if he will not admit thus much, he is putting himself in a radically unreasonable position.

3. That these preliminary remarks, these borrowings from a department of philosophy outside our own, are not uncalled for, will be recognized immediately by any one who will consider the vast difference between certitude viewed from the point of pure phenomenalism, and certitude seen from the point of view here enforced. Of course, as a matter of fact, no one is consistently a pure phenomenalist, believing only in appearances without a reality: and Mill's admission[5] that he cannot regard mind as "a series of states aware of itself as a series," without any bond of union, is a shabby acknowledgment of substance. Nevertheless, the principles of pure phenomenalism are ever being insisted on, to the active promotion of the cause of scepticism; and the perpetual ridicule cast on faculties, or on anything beyond ideas, their associations, and their sequences, necessarily fosters agnostic conclusions. The conclusions, when reached, contradict the principles which have been used to establish them; for, bad as the account is, the account which the pure empiricist gives of the genesis of mind, without substance and without efficient causality, by the heaped-up experiences of unconscious nerve-shock, involves more of real mind in its arguments than ever could have been supplied by a mind so generated. Some real psychological knowledge, and some acute pieces of reasoning, are mixed up with the unreasonable parts of the procedure. The upshot of the whole, however, is logically a complete

---

[5] *Examination*, c. xii. p. 213. See still more what he admits in the Appendix on this subject.

destruction of the edifice of human knowledge. Accept this theory of mind, and you have no mind left.

Therefore, in this treatise, so much stress is laid upon starting from the notions of substance and efficient causality, as from real, indispensable groundworks for a philosophy of certitude. Those who know something of the state of philosophic opinion in this country, will be ready to admit the relevancy of our brief reference to substance and causality, outside of the treatise in which they are properly discussed; and those whose reading has not qualified them to be judges on the matter, will do well to accept our assertion on faith for the present, and verify it themselves hereafter.

R

## CHAPTER II.

### THE TRUSTWORTHINESS OF THE SENSES.

*Synopsis.*
  (A) Preliminaries.
    1. How, as a fact, ordinary people come to believe in their own and other bodies, and in the sensible properties of both.
    2. The universal tendency so to believe in the reports of the senses is a strong presumption for the validity of the belief; but the matter must be argued out in form.
    3. Some distinctions and divisions useful in the course of the argument. (*a*) The number of the senses, and recent discoveries as to the action of the senses. (*b*) Division of the objects of sense. (*c*) Distinction between sensation and perception.
  (B) Proof.
    4. We start the proof from the admitted community of experiences between our adversaries and ourselves as to the sensible world.
    5. Then the trustworthiness of a man's senses is proved; for (*a*) that they testify to the *existence* of his own body and of other bodies is shown (i.) by the admitted existence of "other men," (ii.) by an analysis of the facts of sense-perception, (iii.) by confirmatory considerations drawn from science: and (*b*) that they testify something as to the *nature* of these bodies is also a demonstrable fact.
    6. Summary of the long argument.
*Addenda.*

(A) IT is admitted with tolerable unanimity that the acquisition of knowledge is a process, beginning with the senses; and, therefore, with an examination of their testimony we must start our critical investigation of certitude in detail. During the performance of this task it will be made apparent, how much we need the ideas of substance and efficient causality, and how little we could do, if we were to accept Professor Clifford's dictum, that "the word *cause* has no legitimate place in the science of philosophy;" or the saying of Reid, that "for anything we can prove to the contrary, the connexion between impression and sensation may be arbitrary," and that "causes have no proper efficiency as far as we know;" or lastly, the words of Professor Green,[1] "The greatest writer must fall into confusion when he brings under the conceptions of *cause* and *substance* the self-conscious thought which is their source; when, in Kantian language, he brings the source of the categories under the categories": for "the mind is not *substance*, but *subject*," in which "tersely put formula Hegel emphasizes his position towards the ordinary metaphysics." Such doctrines are absolutely fatal to the claim that man can gain real knowledge through the media of his senses.

1. The philosophical discussion of the validity of the senses may be aptly prefaced by a statement as to what is the way, and the highly reasonable way, in which ordinary people, through their senses, come

[1] *Introduction to Hume*, § 129, § 132. Compare Kuno Fischer's account of this same doctrine, which forms so important a part in Kantian philosophy: "Causality is not the *product*, but the *condition* of experience: it is not *experienced*, but *makes experience*. With regard to the categories, this is the difference between Kant and Hume—between criticism and scepticism." (*Fischer on Kant's Critick*, c. iii. § vi. p. 89, Mahaffy's Translation.)

to the recognition of an external world of matter, distinct from their own bodies. Apart from all philosophy, it is a commonly admitted truth—which the idealist also allows when he is not idealizing, and still allows when he is idealizing, but in his own perverse way—that each man has a body with a set of separate bodily senses attached; and that thus constituted the individual is placed in a world made up of things, which also are bodies. From earliest infancy, all through the long ceaseless course of education, which the senses have to undergo before they become fitting instruments of perception, and thenceforth continuously up to the end of healthy existence, man is ever receiving experiences which go to enforce the conclusion, that there is a thing which is his own body, and that, distinct from this, there are other bodies. Constant action and reaction between organism and environment, as also between different parts of the organism itself, serve to impress this conviction. Daily more and more is the reason satisfied that it is rightly interpreting the situation. It may be that no deliberate, explicitly designed line of argument is gone through: or that if such argument be explicitly attempted, it seems a failure, only obscuring what before was clear. This fact leads a number of writers to say, not accurately, that belief in an external world is not a rational process, that reason destroys natural conviction, and that only instinct is to be trusted. It is more satisfactory as a theory, and more in accordance with the truth of facts, to hold that while no mere verbal argument can contain the full cogency of proof, which is found in a life spent literally in knocking about the world and in being knocked about by it—a life of thumps and bumps against hard matter; yet the argument is capable of verbal expression, in a form which meets the requirement of demonstration. The verbal form is not as forcible as the accumulated experience, but it is argumentatively valid, especially as it is addressed to those who have the experience. From the first tumbles of a child learning to walk, up to the last stumbles of an old man tottering at the verge of his grave, there is, first of all, strong non-philosophical proof that there is solid matter in and out of the human frame. Afterwards the non-philosophical proof can take philosophic shape: in which transformation philosophy has nothing to rely upon except its power to give systematic shape to nature's spontaneous interpretation of experience.

2. That the common, spontaneous belief of mankind is what it is, affords strong presumption that it is right. Clifford, indeed, tries to cast doubt on the fact that the popular belief in an outer world is such as we assume it to be, but herein he is certainly wrong. So is Mill when he declares that apart from philosophic and theologic bias, his view contains all that mankind really believe. In point of fact the common persuasion is, that we have each a material organism, brought into varied contact with distinctly other matter: and in making this interpretation of the case the common voice, as we now wish to argue, is likely to be correct. For the belief concerns not abstruse, remote speculations, but

one of the most fundamental, indispensable notions about the constitution of self-conscious human nature and of its surroundings. Assuredly the presumption is, that the easy, ready, and universal account rendered by our intelligent nature of itself, is better than the strained effort after theory, which, perhaps, its very advocates do not practically believe. Even Fichte himself confessed, that while idealism was, as he fancied, demonstrable, yet it would never be believed.

3. However, we must go beyond presumptions in favour of our thesis, and set about the solid business of proof; for which the way must first be prepared by a few divisions and distinctions, that throw light on the whole matter in hand.

We may leave alone the division of the senses into inner and outer, which raises the controversy whether the seat of all sensation alike is the brain, or whether the outer organs are likewise seats of sensation. Nevertheless, as we are going to treat principally of what are called "the outer senses," we shall do well to frame some answer to the question, how many these are, and how far has the old account of them been upset by modern physiology?

(*a*) To the traditional five senses modern writers make additions by splitting up what used to be comprised under the one faculty of *Touch* into several senses. The resulting new terms have now grown pretty familiar to a reading public that must have been sufficiently often brought across such words as "the muscular sense," and "the sensations of organic life." It has heard also of special nerves, or special conditions of nerve, for perceiving heat: it knows of such curious facts as are implied in analgesia, or insensibility to pain, while there is no accompanying anæsthesia or insensibility to touch. A patient has seen the lancet approach the flesh, has felt the incision, and has wondered at the absence of suffering. Rarely there seems to be anæsthesia without analgesia. These facts are worth knowing; and any one who, treating of the validity of the senses, utterly ignored such discoveries might be suspected of incompetency. But really, on careful consideration it will appear, that with the exception of the stress laid on what is called the muscular sense for coming to the knowledge of resistance, of externality, of magnitude, and the like, few of the new ideas enter much into the present dispute. How for instance does it affect our problem, to be told that the rate of propagation in the nerve stimulus is rather slow, and that, on a rough estimate, while stimulus increases in geometric progression, sensibility increases only in arithmetic? For our business, then, it is enough to have examined what is the style of modern discoveries with regard to the outer senses, in order to assure ourselves that these discoveries offer no obstacle to the arguments we are about to use, and then to decide that the old division into five senses will satisfy our requirements well enough, if we only remember that the division is not very precise. But the general fact itself, that there are different senses is a consideration of some weight in our problem;

because it raises, for example, such questions as, how can these diverse senses be all true reporters, which report so differently of the same object?

(*b*) Next to the division of the senses comes a call for a division of the objects of sense; to meet which demand, obviously one way would be to let the first division settle the second. But there is another division which suggests itself to nearly every investigator, and is often introduced into the controversy upon which we are preparing to enter. For the distinction readily occurs, according to which some sensations are specially referred to the object felt, others specially to the subject feeling, and others not specially to either. The size of an apple, its taste, and the combined feeling of pressure and resistance to which it gives rise when the hand is placed upon it, are instances respectively of the three modes of sensitive experience.

Let us go back to Aristotle,[2] who distinguished with pretty much the same result as the above, those sensibles which can be reached by more than one sense—τὰ κοινὰ αἰσθητά—and those which can be reached by only one sense—τὰ ἴδια αἰσθητά. St. Thomas[3] calls the former *sensibilia communia* and the latter *sensibilia propria*. Thus, at least, in the educated condition of the senses, superficial extension is perceptible both to sight and touch, and is regarded as specially objective; colour, sound, odour, are each perceptible only by one sense, and are regarded as specially subjective. St. Thomas adds a third class, the "Things which fall accidentally

[2] *De Anima*, II. vi.     [3] *Summa*, 1a, q. xvii. a. ii. c.

under the senses, as when this coloured object happens moreover to be a man."[4] Aristotle's parallel instance is seeing the son of Cleon. We see an object of definite colour, light and shade, outline; we know this to be a man, and even to be the son of a certain father: but these latter facts are not at the moment immediate objects of our sight; they are known *aliunde*. The corresponding classification in favour among English philosophers is that according to primary and secondary qualities; or as Hamilton puts it, into primary, secondary, and secundo-primary. He enters into great minutiæ, but we need not follow him. It is enough to have called attention to the fact, that whereas sensation includes an objective and a subjective side, sometimes our attention is called predominantly to the one, sometimes to the other, and sometimes neither side seems to predominate.

(*c*) Hamilton again distinguishes between sensation and perception. Those who push this distinction to the uttermost, describe sensation itself as mere subjective feeling, with no object to which it points, or as not a cognitive act.[5] They make all perception a separate act, supervening on sensation; and they make it the business of the mind to trace this subjective state to some outer cause, almost as

---

[4] "Sensibilia per accidens, sicut quando huic colorato accidit esse hominem." (l.c.)

[5] For example, Lotze: "That which takes place in us immediately under the influence of an external stimulus, the sensation or feeling, is in itself nothing but a state of our consciousness, a mood of ourselves;" it belongs to the activity of thought to convert this "impression" into an "idea." (*Logic*, pp. 10, 11.)

we might interpret the meaning of a foot-print in the sand, saying that it is the mark of an extinct animal. Reid only too manifestly tends to this extreme view and is therefore reprehended by Hamilton. He even goes further and almost leaves the work of assigning the objective origin of sensation in the hands of the Creator. Regarding the perception as an act only of the mind, Reid connects it with the sensation as with a mere antecedent, which may have no closer tie with the perception than the will of God, who has settled that, in fact, after a bodily impression, a mental expression shall follow.

It pertains to psychology to treat this matter, but we may state a few leading heads of doctrine. First of all, sensation itself is something neither purely mental, nor purely material. It belongs, as Hamilton says, to the animated organism, or to united soul and body; the proof of its compound nature being apparent in the felt phenomena, which are partly of a spiritual partly of a bodily character. This composite character of our sensations is of great importance in accounting for our notion of Space, which pure empiricists vainly seek to derive from non-spatial feelings, while the *a priori* school make it a subjective form of our faculty, which they call objective because all men alike have this form. As to whether, besides sensation, there is such a thing as sensitive perception, the condition of the lower animals, is an argument that there is. The Duke of Argyll appeals to our own experience in the matter as very convincing: but, while it is true that we have sense-perceptions, it is also true that we cannot begin reflectively to analyze them except in terms of intellectual perception. What is called the sense-perception of an object is often really the intellectual perception consequent on the sense-perception.

(B) Now what, in the coming argument, we must chiefly have regard to, is precisely the intellectual perception and judgment about objects of which we are made cognisant through the medium of the senses. When intellectually we judge that there is an outer material world, having really such and such properties, then we have the act which this chapter is concerned to prove generically valid. We do not suppose outer objects immediately setting a seal upon the spiritual mind: and Ferrier is quite misconceiving our problem, when under the wrong notion just repudiated, he declares, "Descartes saw that things and the senses could no more transmit cognitions to the mind than a man can transmit to a beggar a guinea which he has not got."[6] We, too, see and confess as much: but what we deem still worthy to be examined into is, whether the intellect can arrive at judgments about the external world, because this world first acts on an animated organism adapted to feel and sensitively to perceive it; and because, on the occurrence of the sensitive perception, the intellect, which is only another activity of the same soul that takes part in the sensation,

[6] Descartes is not uniform in his doctrine about the senses; but he has made distinct admissions that our theory need not imply anything like the literal transference of an image from sense to intellect. See a quotation in Mr. Huxley's *Hume*, p. 84.

is adapted to form to itself ideas corresponding to the objects which excite its sensibility. Undoubtedly it is a very obscure point how the transition is made from sense to intellect; but, as we have to repeat so often, a fact may become apparent while its mode remains undiscoverable. The mode even of the mere sense-reaction has its obscurities, under cover of which some speak as though the re-agency were merely mechanical, and not the re-agency of a faculty, which, in its own lower order, is cognitive. Yet surely a sense-impression is not received simply like a stamp upon wax or a stroke on a bell. The proper attitude under obscurities is neither to deny ascertainable facts, nor to assert as facts what are fictions.

The above divisions and distinctions, even though seldom explicitly appealed to, are most valuable in shedding light on the matter about which we have now to argue; and the absence of them leaves a great haziness of mind, anything but conducive to the work of framing or appreciating arguments.

4. Briefly stated, the whole proof of the present thesis will consist in showing that the experienced facts of sensation are confessedly alike with our adversaries and ourselves, and that only our way of accounting for them is adequate. In other words, starting from the common ground of an admittedly double series in our sensations, we have next to show that the true account of the fact is what has been broadly expressed by the terms *realism* or *dualism*, which mean that there are *two real* divisions of things, "my body," and "bodies outside mine."

Let us start with the declaration of what is common ground.

It would be very awkward, indeed, for us, if we found adversaries asserting that they have no experiences answering to our own; that outer and inner objects, the different personal pronouns, *I*, *you*, and *they*, are terms which correspond to no distinctions in their consciousness. But it is the very complaint of the idealist that his admissions on these points are not recognized, and that he is supposed to be logically committed to an utter disregard of mad dogs, infuriated bulls, express trains, yawning abysses, on the one side; and on the other side, of good dinners, elegant dress, commodious lodgings, and entertaining company. His protest is that all ordinary forms of speech have a meaning for him. He allows that the sun, on present calculation, is about ninety millions of miles off; he expects in about a week to complete a voyage to America and find "the big continent" at the end. He would correct a child who mixed up the doings of Napoleon and of Wellington, and he claims to himself the exploits of neither: he does not at all allow that they are the fictions of his own fancy. Perhaps he will go so far as to talk of a time a long way back in the process of evolution, when consciousness as yet was not. Mr. Spencer thinks the idealist has no right so to speak, Mr. Sully thinks he has, our view of the matter may be given later: at present let us turn to some examples in proof of the unanimity between idealists and realists as to the facts of experience for which an account has to be given. Of course only the idealists need be quoted.

Berkeley,[7] remarking that he can call up fantastic images as he likes, adds, " but when in broad daylight I open my eyes, it is not in my power to choose whether I shall see or not, or to determine what particular objects shall present themselves to my view." "The ideas of sense are more strong, lively, and distinct than those of the imagination. They have a liveliness, a steadiness, order, and coherence, and are not excited at random, as those which are the effects of human wills often are, but in regular train and series." Berkeley, it is true, was only a half-hearted idealist, though, as his notebook shows, he had thoughts of abolishing spiritual substance among created things, just as he abolished material substance, and then he would have become wholly an idealist. If, however, we want a man who, according to his principles, ought to be the most out-and-out idealist, we have Berkeley's continuator, Hume: and he fully admits the contrast between the actual and the imaginary in our objects of thought. "Nature, by an absolute and uncontrollable necessity, has determined us to judge as well as to breathe and feel; nor can we any more forbear viewing certain objects in a stronger and fuller light upon account of their customary connexion with a present impression, than we can hinder ourselves from thinking as long as we are awake, or seeing the surrounding bodies when we turn our eyes towards them in broad sunshine.

[7] *The Principles of Human Knowledge*, nn. 28—31.

Whoever has undertaken to refute the cavils of this total scepticism has really disputed without an antagonist, and endeavoured, by argument, to establish a faculty which nature has antecedently implanted in the mind and rendered unavoidable."[8] Passing on to a great modern representative of Hume, we find Mill[9] owning to an experience like ours, as we gather from what he says about his belief in the permanent existence of icebergs, of a piece of white paper, and of the city of Calcutta. Elsewhere he distinctly recognizes his own bodily senses as the organs whereby he communicates with the external world. " Physical objects are, of course, known to us through the senses. By these channels, and not otherwise, we learn whatever we do learn concerning them. Without the senses we know no more of what they are than the senses tell us. Thus much, in the obvious meaning of the words, is denied by no one, though there are thinkers who prefer to express their meaning in other language." The twin philosopher with Mill, namely, Mr. Bain,[10] though he declares the question whether there is an outer world not to be even intelligible, yet clearly recognizes the experiences which we call those of the outer world: "The perception of matter points to a fundamental distinction in our experience. We are in one condition or attitude

---

[8] *Human Nature*, Bk. I. Part IV. § 1. As Hume wished to be judged by his later work, we may say that similar confessions are found in the *Inquiry*.

[9] *Examination*, c. ix. p. 127 ; c. xi. pp. 192, 199.

[10] *Mental Science*, Bk. II. c. vii. pp. 198—202.

of mind when surveying a tree or a mountain, and in a totally different condition or attitude when luxuriating in warmth or suffering from a toothache. The difference here indicates the greatest contrast." And again: "Object means (*a*) what calls our muscular energies into play as opposed to passive feelings; (*b*) the uniform connexion of definite feelings with definite energies, as opposed to feelings unconnected with energies: (*c*) what affects all minds alike, as opposed to what varies in different minds. . . . The greatest antithesis among the phenomena of our mental constitution is the antithesis between the active and passive." A more appropriate quotation still may be given from the same chapter: To say that the perception of matter is an ultimate, indivisible, simple fact "is as doubtful in itself as it is at variance with the common belief. When we turn to the fact called perception, we cannot help being struck with the *appearance* at least of complexity. There is seemingly a combination of a perceiving mind, a mode of activity of that mind, a something to be perceived—nothing less than the whole extended universe. To make out this seemingly threefold concurrence to be an indivisible fact, would at least demand a justifying explanation." Lastly, to quote the testimony of a prominent scientific man, who more than the common run of his brethren claims to be likewise a philosopher, Mr. Huxley admits that the realistic hypothesis so well satisfies the facts of the case that it may be true:[11] "there may be a real some-

---

[11] Huxley's *Hume*, c. iii. p. 81.

thing which is the cause of our experience." This admission he unfortunately follows up by another admission, which shows the abyss of the agnosticism into which he has fallen, and to which we shall have repeatedly to recur afterwards, because it is such a clear declaration of his philosophical bankruptcy. "For any demonstration that can be given to the contrary effect, the collection of perceptions which makes up our consciousness may be an orderly phantasmagoria, generated by the Ego unfolding its successive scenes on the background of the abyss of nothingness; as a firework, which is but cunningly arranged combustibles, grows from a spark into a coruscation, and from a coruscation into figures and words and cascades of devouring flames, and then vanishes into the darkness of night."

This last avowal is not satisfactory: but at any rate we have the satisfactory result of finding a common account of the phenomena to be explained; and we may now go on to find proof of the manifest breakdown of the idealistic theory and of the manifest stability of the moderate realistic doctrine, when each respectively is called upon to explain the universally admitted facts of experience.

5. It is not with the whole of idealism that we have got to do, but only with the part which concerns the sensible world of matter. However, the fundamental difficulty, on which throughout idealism is based, is contained in the question, how can the individual get outside of itself? how can thought transcend itself? how can the subject know any-

S

thing except its own affections? [12] In reply we have to repeat the old truths, that we may be certain of a fact without being acquainted with the how of the fact; and that "from a fact to its possibility the inference is valid." [13] At least it is a piece of more gratuitous dogmatism than they seem to be aware of, when idealists lay it down *a priori*, that it is a plain self-contradiction to suppose the perception of an object, which object is other than the percipient, and known by him to be such. Not that there is no mystery in the process: indeed there is mystery even in the simplest instance of what we call a transient action, as when a moving body sets in motion a body before at rest. Still more is there mystery in the process of thought, an act at once physically immanent in the subject, and transient, as the scholastics say, *intentionaliter*, that is, having its term, so far as meaning and intelligence are concerned, something outside the subject. The mystery then we allow: but at the same time we contend, that however mysterious, still a fact which can be established ought to be recognized. In order to the establishment of the fact we have two points to prove: (*a*) that each one's senses testify to the

[12] See Mr. Bain's *Mental Science*, Bk. II. c. vii. p. 198. "The prevailing doctrine is, that a tree is something in itself, apart from all perception; that by its luminous emanations it impresses our mind and is then perceived; the perception being an effect, and the impressing tree a [partial] cause. But the tree is known only through perception, we can think of it as perceived, but not as unperceived. There is a manifest contradiction in the supposition; at the same moment we are required to perceive and not perceive."
[13] "Ab esse ad posse valet illatio."

*existence* of his own body and of bodies not his own; and (*b*) that they testify something about the *nature* of these bodies.

(*a*) In behalf of the former point three arguments may be adduced.

(i.) Our adversaries each assert *the existence of other men*, and it is on this ground that we will do battle with them in the first instance. Relegating all account of individual writers to a note in the *Addenda*, lest it should here perplex the course of an argument already sufficiently difficult in itself, we must be content to speak in quite general terms. We say, then, that on the strength of sensible manifestations, opponents are quite unwarranted in their inference that "other men" besides themselves exist. By the very principles of their position they are shut out from the conclusion that anything is truly other than their own sensations; and their pretence that "other men" are demonstrable while "external matter" is indemonstrable, can be kept up only by a delusion resting on great confusion of thought. For in the end it will be seen that the assumption of a known "external matter" is needful, and is employed in the argument whereby the conclusion is drawn that there are "other men." A reference to Mill's view, as explained in the *Addenda*, will make this point clear. The strength of our attack on the adversaries always lies in this: they assert distinctly "other men" with bodies like their own, and thereby they give up their own doctrine as to the power of the senses.

After showing the inability of idealists to defend

their belief in "other men," we may now venture upon doing what they have failed to do, framing upon their suggestion an argument of our own, which, while it is not one ordinarily used in books, is an effective demonstration of the validity of the senses. The line of proof runs thus. We certainly do, through our senses and the material manifestations furnished to them by "other men," come to a sure knowledge that these men exist. But this could not be, unless our senses were valid means for reaching the knowledge of external bodies. Therefore our senses are such valid means. The major of the syllogism can be established in a special way, which will leave untouched the commoner arguments that are to be adduced presently. For, that we do come across other minds, is most clearly evidenced to us by the intellectual assistance we receive from them. It would require a very foolish or a very shameless scholar, seriously to maintain that all the information he receives from teachers and books is really as much the exclusive product of his own mind, as that which he ordinarily calls his original thought or discovery; allowing this sole difference, that the former knowledge is accompanied by a special feeling of derivation from outside, which is, after all, only a part of his own inner consciousness. Let us think of our very, very wide indebtedness to other minds; how very much less than we are, we should have been, intellectually, had others not taught us orally or in writing; how very little we really know at first hand: and then let us try to swallow down, we might almost have called it the idealist joke on the subject, were it not that some idealists are manifestly in earnest. We feel that we have not powers of deglutition for so formidable a morsel. If then we really do come in contact with other minds, and draw knowledge from them, the intercommunion is certainly not one purely spiritual: it is through the senses and by means evidently material. With our bodily senses we approach those bodily objects, the books of the British Museum, the Natural History Specimens in its Kensington offshoot, the libraries, the custodians, and the professors, who, as experts, help us inexperts out of many a difficulty. Surely the least recognition we can pay to our kindly helpers is to acknowledge unreservedly their real, independent existence. Mr. Huxley, in spite of his theory that idealism cannot be disproved, expresses himself gratified with the tokens of esteem that he receives from former pupils. Now if he would good-naturedly consider the impossibility of his harbouring any genuine doubt, as to whether he has been exercising and receiving the offices of real "altruism," or has simply been teaching himself under another form, and receiving from the pseudo-outsider compliments, which his modesty would have forbidden him undisguisedly to pay to himself; he might be brought to recognize that the existence and the actions of really "other men" can be fully brought home as a conviction of the reason, and that idealism, in consequence, is exploded, not only practically, but theoretically. He would retract the already quoted passage, that for aught we can demonstrate to the contrary, all our thinking may be so many idle fireworks let off by the mind against "a background of nothingness."

Beyond a doubt, under the single category of the intellectual aids which we derive by our communication, through the senses, with our fellow-men, there lies proof positive that idealism is an insulting attempt to fool a man out of those faculties which are his birth-right. Because we are treating philosophically of the senses, we are not therefore to allow ourselves to be staggered "out of our five wits," by any phantom which a bit of sophistry may conjure up before us. Because we have on the philosophic mantle, we are not, therefore, to yield up that sound judgment which we possess, when we are, so to speak, in our shirt sleeves. In the latter condition we are ready to fight a pretty vigorous battle for the reasonableness of trusting our senses; and there is nothing to prevent us, as philosophers, from doing the same stout battle. As philosophers we may affirm, what as ordinary men we affirm, that there is evidence from the senses, such as to warrant our belief in the existence of our fellow-mortals; and that in this conclusion is involved the wider proposition, that about the world of matter in general our senses can testify to its outer reality.

(ii.) To pass now from the consideration of "other men," a consideration which our adversaries have usefully forced upon us, we may turn to the arguments more commonly adduced on behalf of the senses by standard authors.[14] Each one who is unburdened by Kantian views as to space and time, may formulate to himself this argument in some such shape as the following: I can verify for myself, as an explorer, the existence of my own extended body, of definite shape and size. At least by repetition and comparison of experiences from different senses, I can become aware of my several sentient organs; of one sensation as being peculiar to one inlet, and another to another; of sights entering in at places different from the places where sounds enter. I can feel the double sense of contact, that of touching and being touched, when I place my right hand upon my left, and I can contrast this duplex sensation with the single sensation given by putting either hand upon the table. Gradually, if not at once, I can explore the limits of my sentient body. I find this body of mine at the same time brought into relation with other bodies, in such sort that the only rational interpretation of the situation is to say, these bodies are really not mine. I touch them and feel their resistance to my energies, but invariably without the double sense of touch or resistance which I usually have when it is one part against another part of my own body that I oppose. Conviction is, in a million instances, brought home to me that I am passively sentient, not of course with a pure passivity, under many outside influences—influences which I cannot have at will, or carry about with me,

---

[14] Tongiorgi, *Logica*, Part II. Lib. II. cap. iii.; *Logik und Erkenntnisstheorie*, von Dr. C. Gutberlet, Zweites Kapitel, pp. 174, seq.

or vary with the same degree of control which I have over a mere train of subjectively originated imaginations. The control in the latter case is indeed far from absolute, but at least it is perceptibly something. Nor can I persuade myself, on Hume's suggestion, to get over the difference between real and imaginary objects by attributing it only to a greater and less degree of subjective liveliness; for I have the means, while reason lasts, of detecting even very lively fantasies to be only fantasies.

So might a common man argue, and validly. It is because he so reasons that he is apt to receive the often inculcated lesson of scientific men, like Mr. Huxley, that about physical facts we must consult outer nature, and not try to evolve them from our inner consciousness. If we want personally to explore the home habits of the Polar bear, we must join a Polar expedition, which will mean a great deal more than the idea of a tedious and perilous voyage preceding the idea of finding what we seek. Yet according to strict idealists this is all that is meant. For instance, Professor Huxley says [15] that the analysis of the proposition, "Brain produces thought," "amounts to the following: whenever those states of consciousness which are called sensation, motion, or thought, come into existence, complete investigation will show good reason for the belief that they are preceded by those other phenomena of consciousness to which we gave the names of matter and motion." As the Professor cannot mean that we always think of matter and

[15] Huxley's *Hume*, c. iii. pp. 80, 81.

motion before we think of consciousness, he has no right to call the cerebral motion which, on the theory of brain producing thought, would be the antecedent of consciousness, by the name of a "phenomenon of consciousness." How can that antecedent be the phenomenal antecedent in consciousness which in consciousness does not antecede the result?

The main difficulty brought against this, which we have styled "the ordinary argument" for realism, is made to rest on impossible theories about the origin of the notion *extension* or *outness*. It is asserted that *local outness* is not given simply by the consciousness of one thought being other than a preceding thought, and then great labour is expended to develop externality in space out of succession in sentient states. These bugbears set up by a bad psychology must be encountered in the psychological treatise; but we in our own treatise at least are justified in claiming, on the strength of natural evidence, a clear idea of outness in space as derived through our sensitive experience. We need no more for the purposes of the line of proof just brought to an end.

(iii.) It is not necessary to develop further the argument against idealism and for realism as furnishing the genuine account of those experienced differences between inner and outer bodies, which all parties admit, but some confirmation of what has been urged may be borrowed from Professor Tait's idea, that the great proof of external reality is the scientific truth that matter can neither be created nor annihilated. On idealist principles this proposition might still be held, but it would have very little value. As soon as the scientific man was persuaded that matter was only the objective side of his ideas, without ascertainable independent existence, he would care very little about its increase or decrease: and might even claim to increase and decrease it at will, at least under certain conditions.

Another confirmation, suggested by Mr. Spencer, and allowed by Mr. Balfour, but disallowed by Mr. Sully, lies in the assertion, that "if idealism is true, then evolution is a dream." For evolution supposes an indefinitely long period, during which there was no consciousness in the universe. Such a universe, as an existence, cannot have been ideal, and cannot be affirmed now by the idealist: for it would once have been a universe out of all human thought, which Mr. Bain, on his principles, rightly concludes to be a "manifest contradiction."

(*b*) Some, conceding to us all which so far we have been pressing to prove, but not all we have actually proved, would bid us stop short here; they admit that we have evidence for predicating the bare of *existence* of bodies outside our own, but nothing more; we can say nothing of their *attributes* or *nature*. Kant, in some passages, but not in all, takes up exactly this position, and Schopenhauer declares "he must be abandoned by all the gods who imagines that there exists outside of us a real world of objects corresponding to our representations."

At this juncture the distinction is of some use between what are called **primary and secondary** qualities, though it is not to be pushed to excess, as though any sensible quality could be perceived as quite out of all relation to sense. We may contrast the relations we affirm between the object and the organism of the subject, with the relations we affirm between one object and another. Whether sugar is sweet, ginger hot, and aloes bitter, depends upon the subject, and would change with a possible change of subject; but no change of the subject's faculties could validly report that St. Paul's would go inside the smallest shop in Paternoster Row, and that a strip of carpet, which we have in a corner of the room, would cover the whole floor. It is true enough that all objects, whether primary or secondary qualities, affect our senses relatively to the structure of our organs; but not only can there be no knowledge of relations without some knowledge of the absolute terms which are related, but in asserting one class of relations between external bodies, we assert that which would not change with a change of our organism, though this latter change might increase or decrease our perception of the outer facts. That a whale is larger than a whiting does not depend on any percipient organism, but is true for any organism that can perceive it.

Again, when we think of some well-established chemical analysis, for example, the resolution of water into two gases, we ask ourselves, is there no real insight into the nature of things here? Is physical science so devoid of objective reality as to tell us nothing of "things themselves," in the rational meaning of that phrase? Is the resistance we directly encounter from external objects nothing proper to the objects themselves? Is it a fact that we can regard it only under the false analogy of a will-power, never as a material power? It is suicidal in the idealist to quote, as he does, the instances of light and heat, and to argue his case with an air of triumph, from the fact that vibrations of a fluid medium are quite unlike the sensations of sight and hearing. He forgets that it has been by the senses that the vibrations have been discovered, and that if the scientific result is worth anything, it proves the ability of the senses to give us information about facts as they are in external nature. To urge in reply that these facts are, for us, only as known by us, not as existing out of relation to all knowledge, is futile; for this does not prove that we cannot know objects as they really are. We do not know all about them, but that we never claimed to know; at least we know something, and that contradicts idealism.

In saying that our knowledge is a compound of subjective and objective elements inextricably combined, adversaries make the mistake of going simply on the analogy of a chemical composition.[16] Water

[16] Kantians sometimes speak in this sense, and sometimes they make the whole perception subjective. "The *external object*, or what we call *the thing without us*, is not by any means the thing *per se*. The thing without us, resolved into its elements, consists of sensation and intuition, partly our datum and partly our product: it is nothing but our phenomenon, our representation. The thing *per se* is a term by which we designate the very opposite of this, namely what can never be phenomenon or representation." (*Fischer on Kant's Critick*, pp. 53, 54.)

is neither oxygen nor hydrogen, being a chemical compound of the two. But thought is not a chemical compound, having for its constituents object and subject. Materially the known object has not to be shot into the mind and fused with it. The reaction of mind after the stimulation of the senses, is not any kind of a reaction, but a definite, most peculiar, and most exalted one. And the argument which urges that no knowledge attains to reality as it is, because all is relative, is so radically false, that it includes not only finite minds, but all minds, even the Divine, and denies to God Himself an absolute knowledge. Its perspicacious and consistent advocates boldly affirm, that from its very nature no knowledge can be absolute, attaining to the thing as it is; knowledge must be relative, must transfigure its object, must mix up elements or forms of self with elements or forms of non-self. No such *a priori* reasonings are valid. There is no demonstration that even a finite faculty must so transfer its own conditions to objects as known by it, that it can know nothing rightly. The only point demonstrated is, that a finite faculty will have many limitations, because of its imperfection; but that knowledge, as such, cannot in any intellect be absolute and complete, is the merest piece of perverse dogmatism, without the shadow of a proof. Lay bare the falseness of an analogy between knowledge and chemical combination, and all argument for the dogma collapses.

Let us end with an illustration from one of the primary qualities of body, impenetrability. A poor prisoner in Newgate does not beat idly against the walls of his cell, like a bird just caged. For intellectually he perceives that huge blocks of masonry are hopeless obstacles; that they bar the progress of a man who would walk through them. Immoveably they occupy the space where they now are, and in the fact that two different material bodies cannot naturally[17] occupy together identically the same space, consists the familiar property of impenetrability. So thinks the prisoner. But Mr. Huxley, who is at large in the world, solemnly tells it, that, "if I say that impenetrability is a property of matter, all that I can really mean is, that the consciousness I call extension and the consciousness I call resistance, inevitably accompany one other." We cannot think of impenetrability without consciousness; but all the same we can know impenetrability to be a real property found in unconscious matter, and belonging to it, not because of our consciousness.

While maintaining that our senses enable us to form some correct judgments about matter and its properties, we fully admit how far from exhaustive is our knowledge. Take for example the properties of extension in space and succession in time. A Catholic least of all would arrogate to himself,

---

[17] We say naturally, because we do not deny that preternaturally two bodies may together be in the same place. Hence it is not wholly true to say that the "otherness" of bodies loses its objective reality, if with Kant we make space not something, in our sense of the word, objective, but a mental form of the subject; for "otherness" radically rests not on difference in space, but on the fact that *this* body individually is not *the other* body.

on these points, a comprehensive acquaintance; for some of the mysteries of his faith warn him to the contrary. He easily admits these to involve no clear impossibilities; for he easily admits his own ignorance, and the possibility of that being brought about preternaturally, which naturally would not be. But he does not, on that account, easily forego his own knowledge of simpler truths about the material universe, so long as matter is left in those normal conditions with which he can familiarize himself.

6. Our argument, which has been long rather than abstruse, calling for patience rather than for extraordinary penetration, may now be summarized. In the phenomena of sense-perception rival schools are substantially agreed about the conscious experiences of which an account has to be rendered. Pure idealists, on their own principles, cannot use sensible manifestations to make certain of the existence of other men like themselves; they assert these "other men," but inconsistently, and at the price of renouncing their theory, and coming over to our side. Contrariwise we realists find a strong argument for our doctrine in finding how enormous is the help we receive from our fellows through the aid of the senses. Again, idealists allow, but do not account for the general contrast between sensations of self and sensations stimulated by bodies outside self: whereas we render a rational interpretation of the antithesis—an interpretation so rational that Mr. Bain himself, writing in *Mind*, can condescend to say: "Every one of us readily admits that our impressions are transient things;

yet they come up again with astonishing regularity in the appropriate situations; *and the easiest way of figuring to ourselves this regularity is to suppose a permanent something, with all its parts well knit together, so as to repeat our conscious state with a fixity that we actually find. This is ordinary realism.*" The scientific doctrine of the constancy of the sum total of matter, and the evolutionary hypothesis, according to which, for a long time, there was no conscious existence in the material universe, are conceptions which are badly in accord with idealism, but intelligible to realism, even when the realist does not believe that all life has been developed by the mere self-organization of dead matter. Moreover, not only have we proof of the existence of our own and other bodies, but likewise it is clear that we know something about their nature and their attributes. It would be to know something, if we could predicate of them only the secondary qualities, as that sugar is an object exciting a sweet taste in the palate, and that vinegar rouses an acid feeling; but we can go further and know the primary and more intellectual qualities; for instance, we know about extended space such truths as geometry teaches, and we know about motion such laws as help to form the science of mechanics. The judgment may at times err in its interpretation of the object which is exciting a sensation, but the senses themselves always report what, under the circumstances, they ought to report; and no sensation, as such, can be false. Under the normal condition of the faculties, there is no sensation which is not, of its own nature, calculated to give some information about the material world. A diseased state of organism may baffle the understanding; but it is beyond cavil that there is a state of organism which is normal, and which we have a right to assume as our standard for testing the validity of the senses. Thus, an examination of the whole case leads to the conclusion, that the common belief in the testimony of the senses is well within the bounds of reasonable procedure; and that, in doing what he cannot help as regards trust in his senses, man is not being driven by a blind instinct, but is acting according to his intelligent nature. The instincts of a blind nature are blind; but the instincts of an intelligent nature may often be shown to be intelligent. It is so with our use of the senses.

### Addenda.

(1) We omitted (with a view to avoiding distraction from the main argument) any details as to the way in which our opponents come to the assertion of "other men" beside themselves; these may now be supplied. The substance of Mill's view is contained in the following passage:[1] "I am aware of a group of Permanent Possibilities of Sensation which I call my body, and which my experience shows to be a universal condition of every part of my thread of consciousness. And I am also aware of a great number of other groups, resembling the one I call my body, but which have no connexion, such as that has, with the remainder of my thread of consciousness. This disposes me to draw an inductive inference,

[1] *Examination*, Appendix, p. 253.

that other groups are connected with other threads of consciousness, as mine is with my own. If the evidence stopped here the inference would be but an hypothesis, reaching only to the inferior degree of inductive evidence called analogy. The evidence, however, does not stop here: for having made the supposition that real feelings, though not experienced by myself, lie behind these phenomena of my own consciousness, which from the resemblance to my own body I call *other human bodies*, I find that my subsequent consciousness presents these very sensations of speech heard, of movements and other outward demeanour seen, and so forth, which being the effects or consequences of actual feeling in my own case, I should expect to follow upon those other hypothetical feelings, if they really existed: and thus the hypothesis is verified. *It is thus proved inductively that there is a sphere beyond my consciousness*, that there are *other consciousnesses* beyond it. There exists no parallel evidence in regard to matter."

Now the fact is, that Mill proves his "other consciousnesses" only on the tacit assumption of "other matter:" and to real *otherness* in either department he can never logically attain. For logically he has no right to pass beyond the limits of subjective idealism. Mr. Balfour[2] is positive in the assertion that "there can be no doubt that Mill considered himself an idealist:" and certainly he succeeded in establishing nothing above an idealistic existence for his "possibilities of sensation," however boldly, after denying the reality of substance and of efficient causality, he might arrogate to his "possibilities" both substance and efficient powers. It is part of the want of clear consistency in the man[3] to account for physical changes

---

[2] *A Defence of Philosophical Doubt*, c. ix. p. 186.
[3] *Logic*, Bk. I. c. iii. §§ 5, 7, 8, 9, et alibi passim.

by "one group of possibilities of sensation modifying another such group," whilst he also taught "that all we are conscious of may be accounted for without supposing that we perceive matter by our senses: and that the notion and belief may have come to us by the laws of our constitution, without their being a revelation of any objective reality:" and that "the *non ego* altogether may be a mode in which the mind represents to itself the possible modifications of the *ego*." Again he asks: "How do I know that magnitude is not exclusively a property of our sensations?" And he holds that we do not know whether, as affirmed of Matter itself, the word divisible has any meaning. Lastly, in controversy with Mr. Spencer,[4] he says: "Neither of us, if I understand Mr. Spencer's opinion aright, believe an attribute to be a real thing possessed of objective existence; we believe it to be *a particular mode of naming our sensations*, or *our expectations of sensation*, when looked at in the relation of an external object which excites them:" yet so that these so-called "exciting objects" must not be considered either as substances, or as efficient causes, or as something really external and independent.

Mill being thus in many ways committed to idealism, cannot argue the existence of "other consciousnesses" or "other men," from the data of their external manifestations: he is wholly shut out from every notion of real "otherness." And yet that his argument does ultimately fall back on the inference of human agents from human activities, other than his own but like his own, will again appear, if we add a concluding specimen of his doctrine.[5] "By what evidence do I know that the walking and speaking figures which I see and hear,

[4] *Logic*, Bk. II. c. ii. § 3, in a note at the end of the paragraph.
[5] *Examination*, c. xii. p. 208.

have sensations and thoughts—in other words, possess minds? I conclude that other beings have feelings like me, because first, they have bodies like me; and secondly, because they exhibit acts and other outward signs, which in my own case I know to be caused by feelings." If Mill had once shown us how he arrived at the *otherness* of the manifestations, we could allow him the *otherness* of the human agents; but *otherness* is wholly denied to his principles.

Perhaps what Professor Clifford says will help to explain why Mill insisted so much on "other consciousnesses," namely, that while "material objects" may be spoken of as "the other side of my consciousness," it is absurd to speak of "other consciousnesses" as only "the other side of my consciousness." To signalize this special character, Clifford calls "other consciousnesses," not objects, but *ejects*, for they must be projected outside of self—"they cannot be a group of my feelings persisting as a group." As to the difficulty of asserting any "otherness" beyond his own thinking self, Clifford thinks he need not waste time over considering a step which his ancestors took for him long ago.

M. Taine avowedly tries to lend a helping hand to Mill for the purpose of securing a little more reality to external objects than his friend's theory can afford. He allows that to us a stone is "a more or less elaborate extract from our sensations;" but further, "we may upon authentic evidence refer to things some of those more or less transformed and reduced materials, *and attribute to such things a distinct existence without us,* analogous to that which they have within. In this respect a stone is a being as real and as complete, as distinct from us, as any particular man. By this addition to the theory of Mill and Bain, we restore to bodies an actual existence, independent of our existence."

It is instructive to see idealists trying in vain to get out of the position called "solipsism," or belief in self alone. Especially they feel that "it is not good for man to be alone," and so they labour strenuously to justify their assertion of "other men" besides themselves; but always with the result of violating their own idealistic principles.

(2) On the subject of primary and secondary qualities of body, Hamilton teaches that we regard objects sometimes "as they are in themselves," sometimes "as they affect us," and sometimes in a half-and-half way: these last qualities he calls secundo-primary. For Hamilton's three terms others substitute mathematical, mechanical, and physiological properties; while Mr. Spencer prefers to use, as almost equivalent terms, statical, dynamical, and stato-dynamical.

(3) Though some regard materialism as the contrary extreme of idealism, Mr. Huxley is constant in his theory that an idealist may be a materialist, though he himself refuses to be either. Let us extend one of the quotations given in the text: "If we analyze the proposition that all mental phenomena are the effects or products of material phenomena, all that materialism means amounts to this, that whenever these states of consciousness which we call sensations, or emotions, or thought, come into existence, complete investigation will show good reason for the belief, that they are preceded by other phenomena, to which we give the names of matter and motion. All material change appears in the long run to be modes of motion; but our knowledge of motion is nothing but that of a change in the places and order of our sensations: just as our knowledge of matter is restricted to those feelings of which we assume it to be the cause."[6] This comes to little more than

[6] Huxley's *Hume*, c. iii. pp. 80, 81.

the jejune announcement, that if matter be reduced to idealistic dimensions then materialism and idealism are reconciled. But how does this square with the evolutionary hypothesis that ideas, for a long time, did not appear, but supervened, in comparatively recent times, on a world of unconscious matter, which cannot be reduced to feelings?

(4) The special form of idealism introduced by Berkeley has so few patrons that it is not necessary to labour much in its refutation. He supposed that all the sensible impressions, which we call material, were due, not to the action of any independent matter, but to the immediate agency of God. With regard to external bodies the difficulty of the theory is somewhat less; but with regard to our own bodies, it would be a task even to Omnipotence to make us feel ourselves as sentient, extended beings, if all the while we were pure spirits, of an essentially unextended nature. Moreover, given such a God as Berkeley rightly admitted, his theory as regards bodies other than our own, is dishonourable to the Creator rather than, as it aims at being, honourable. For an adequate reason, and after a sufficient warning, God may permit such deceptions as may take place through the senses, because of the mystery of the Blessed Eucharist, on the explanation given of it by Catholic theology; but He could not consistently with wisdom and truthfulness, arrange a wholesale system of delusion, such as only a Berkeley here and there would detect, while the mass of mankind were inevitably being duped. Few as have been Berkeley's followers, some of our modern writers in this country have an affinity to him, as, for example, Professors Green and Caird. One of these talks much about finite minds "becoming the vehicle of an eternal complete consciousness," which is " a consciousness operative

throughout our successive acquirements, and realizing itself through them," " an eternal consciousness operative in us to produce the gradual development of our knowledge." These are some of Green's phrases, while Professor Caird's expressions are such as these : " The data of sense are taken out of their mere singularity of feelings and made elements in a universal consciousness : that is, they are related to a consciousness which the individual has not, as a mere individual, but as a universal subject of knowledge. Only in relation to such a consciousness can an individual know himself or any other individual as such." But, perhaps, it is Ferrier who most of all approaches to Berkeley. Ferrier, denying that matter *per se* has any meaning, makes the perception of matter the ultimate, indivisible unit of knowledge. He wholly rejects the analysis into perception as subjective, and matter as objective; he declares the subjective element to be *our apprehension*, that we perceive matter, and the objective element to be *our perception of matter*. Still, he will not allow that the perception of matter is a mere modification of our own minds: he will not lapse into subjective idealism. And it is thus he guards himself against this doctrine: " Our primitive conviction is, that the perception of matter is not, either wholly or in part, a condition of the human soul; is not bounded in any direction by the narrow limits of our intellectual span; but that it 'dwells apart,' a mighty and independent system, a city filled up and upheld by the everlasting God. Who told us that we were placed in a world composed of matter, and not *that we were let down at once into a universe composed of* external perceptions of matter, that were beforehand and from all eternity, and into which we, the creatures of a day, are merely allowed to participate by the gracious Power to whom they really appertain ?

When a man consults his own nature in an impartial spirit, he inevitably finds that his generous belief in the existence of matter, is not a belief in the independent existence of matter *per se*, but is a belief in *the independent existence of the perception of matter, which he is for a time participating in*. The very last thing which he naturally believes in is, that the perception is a state of his own mind, and that the matter is something different from it, and exists apart *in natura rerum*. It is the perception of matter, and not matter *per se*, which is the kind of matter in the independent and permanent existence by which man reposes his belief. This theory of perception is a doctrine of pure intuitionism: it steers clear of all the perplexities of representationism."[7] Ferrier's great point of contention is that matter detached from thought is a delusion; for in pretending to detach it we are all the while thinking about it. It is like pretending to think ourselves annihilated; we find ourselves contemplating the condition; that is, we re-introduce the self we make show of abolishing. It is a simple answer to say, that though we can know matter only so far as it is an object of our ideas, yet we can know that this matter with certain properties has an existence outside our mind. There is no contradiction in the geologist affirming, Had I never discovered it, the fact would still have been, that this rock was scoured and striated by glacial action thousands of years ago.

(5) The very fact of having tried to argue out the validity of the senses is a confession that the result may be reached mediately; but this leaves untouched a further question, whether we have any primarily immediate perception of a material world as external, that is, whether we have any primary intuition of the outness of an object

[7] Ferrier's *Remains*, Vol. II. pp. 454—456.

which we perceive, or whether externality at first can be reached only as a matter of inference. In point of fact, the process of ratiocination is so thoroughly a case of repeated and combined judgments, that the distinction put between the two acts, judgment and ratiocination, by logicians is not so radical as some suppose. We judge and judge again, and put our judgments together, but it is the same intellect which is at work throughout. Now every one will admit that in our present condition of experience we can in some cases immediately judge of externality; and every one will admit that the full reflex distinction between outer and inner world, was not made by the child without several repetitions of acts. So much being settled, we may leave it to psychologists to push further the investigation whether it is necessary to assume an immediate intuition of the externality of the sense-world, or whether the knowledge of this rests on a spontaneous inference as to the origin of some of our bodily affections—an inference so spontaneous that it is taken for immediate perception. All sensations are bodily affections, and the inferential school say that it is only by argument that we can, in some of these affections, detect an outer cause; while the intuitive school declare that this process cannot have begun in argument, without an immediate perception. Outside the sense-world and in relation to metaphysical truths, it is certain that we have immediate intuitions of principles which we at once see to be objective and independent of ourselves; but how the case stands as regards the perception of the outer world of sense, gives rise to dispute among philosophers.

(6) Another psychological difficulty is also involved in our present inquiry. The passage from the image in the sensitive imagination to the idea in the mind is an obscure problem. The mind does not gaze upon the

sensitive representation and consciously copy it. We are safe, however, in affirming, though the affirmation hardly amounts to an explanation, that because of the harmonious working of the faculties in a being whose author is all-skilful, when the sense image is duly present, the intellect has the power to produce its own corresponding image. The harmony is as natural, as certain, and as little ultimately explicable as the correlation of growth in the body, as the adaptation of bodily functions *inter se*, and as any symmetrical arrangement of organic parts; whilst, however, what we call nature has credit for so much, education must step in and take a large share in the formation of our power to perceive by the senses. Our education began so early, and has been so continuous and gradual, that we are apt to overlook the fact. It requires almost a case of congenital cataract cured in later life, to bring home to us the need which the eye has of being trained to do its work. Most of our educated sense-perceptions are such, that what is actually, here and now, presented, is a small fraction of the whole, which is filled up by association or inference. Whatever revelations have been made by Wheatstone's ingenious contrivances for producing ocular illusions, by means of familiar effects under unfamiliar circumstances, all these we must readily acknowledge, without any fear for the truth of our main proposition that the senses are, in their own order, veracious.

(7) There is a deceptiveness about some authors who seem, in places, to agree with our realism, and yet do not. Thus Mr. Spencer argues for realism, and we may adopt some of his arguments. But a further knowledge of his system tells us that he reduces the really distinct phenomena of self and not-self to a basis in "one Unknowable Reality;" and others who do not

explicitly make this final reduction, at least allow its probability. This is called "Monism," the doctrine that all manifestations, however different, are manifestations of but one underlying Entity; and the opposite doctrine is called, with less propriety, Dualism, which means that self and not-self are really distinct existences, the non-self being, of course, a congeries of many existences. The doctrine maintained in this volume is clearly dualistic—an explicit statement which may seem needless. But any one who has had experience of the difficulty of trying to put together all the various declarations of an author, for example, like Lewes, will feel thankful to a writer who will declare undisguisedly where he stands.

(8) Where Monism makes itself most awkwardly felt, is in the distinction between man and man. Probably Mr. Spurgeon does not more strongly feel that he is really not Mr. Huxley, than Mr. Huxley feels that he is not Mr. Spurgeon; and yet, if they are both manifestations of one "ultimate unknowable reality," the identification between them is closer than they might like. As we saw above, those who are idealists, or who admit idealism as possibly true, do not satisfy us that they have sufficiently applied their theory to the distinction between themselves and other men. They are far too apt to assume this distinction, and to argue only for the common nature of the distinct individuals. Thus Professor Clifford says: "I have absolutely no means of perceiving your mind. I judge by analogy that it exists, and the instinct which leads me to come to that conclusion is the social instinct, as it has been formed in me by generations during which men have lived together; and they could not have lived together, unless they had gone upon that supposition." Similarly Mr. Huxley is intent mainly on the analogy

between individuals, not on vindicating, according to his own principles, the real difference between individual and individual: "It is impossible absolutely to prove the presence or absence of consciousness in anything but one's own brain, though by analogy we are justified in assuming its existence in other men." He admits that he cannot be absolutely certain of any "otherness" beyond his own thoughts.

(9) We have taken as our standard the healthy condition of the senses; and without denying to Dr. Maudsley the use of pathological cases, yet we may dissent from the prominence which he gives to them. His professional dealing with so many abnormal specimens of humanity, seems to have given him an unfair opinion of the race in general, or of the average man; and in reading his books it is useful to bear this fact constantly in mind.

# CHAPTER III.

### OBJECTIVITY OF IDEAS, WHETHER SINGULAR OR UNIVERSAL.

*Synopsis.*
1. Proof of the validity of the senses is only a part of the general refutation of idealism; ideas are not mere refined sensations but reach objects above the sensible order.
2. Various forms of idealism.
3. What we have to establish in general.
4. Arguments for this purpose. (*a*) There is no self-contradiction in the way in which the realist supposes thought to transcend itself, and to reach out to objects distinct from itself. (*b*) Idealism is contrary to self-evident truth, and in its extreme form cannot be asserted without refuting itself.
5. Caution against taking too narrow a view of what is meant by the reality of the object.
6. Special difficulty as to the reality of universal ideas. (*a*) The possibility of a finite nature being specifically repeated in many individuals: a repetition which is impossible to an infinite nature. (*b*) Universality is fundamentally in things, formally in the mind alone; hence the determination of the reality proper to a universal idea. (*c*) The insufficiency of the pure sensist view, and of the analogy borrowed from the average photograph. (*d*) The purpose served by multiplying observations and comparisons of individuals in forming the universal idea. (*e*) How we manage to use common terms, which are not perfectly universalized. (*f*) Not at all need we fancy, that every word is one definitely universalized term. (*g*) Difficulty raised against the possibility of abstraction, on the score of inseparable association in experience.
7. Conclusion.

*Addenda.*

1. IT would be an error to limit the problem of idealism to the material world; and hence the last

chapter does not cover the whole of the ground which has to be covered. A question more deep-reaching and more universal is, whether our ideas in general attain to objective reality, be this material or immaterial.

That our ideas are not bounded by our sensations, but have a wider range, must be allowed by all who will take the trouble to go through an analysis of the notions which they possess.[1] It is true that a trace of man's organic conditions clings to his highest intellectual actions; but all the same these clearly manifest a power above sense. Against the theory advocated by Hume, and more or less favoured by many other English philosophers, that ideas are faded, attenuated, and almost etherialized sensations, facts are in dead opposition. Even Lewes, who so largely makes verification by the senses the criterion of real knowledge, has the candour to say: "Ideas are not impressions at all, and hence not faint impressions. Ideas are not sensible pictures. The least experience is sufficient to convince us that we have many ideas which cannot be reduced to any sensible picture." Mr. Huxley, in his manual on *Hume*, is also a witness in our favour, maintaining that "the great merit of Kant is, that he upholds the doctrine of the existence of elements of consciousness which are neither sense-experiences nor any modifications of them." Plain facts of self-analysis do not need the support of confessions made by adversaries; but such

---

[1] Aristotle (*Metaphysics*, Bk. I. c. i.) makes this distinction his very starting-point.

support may usefully be borrowed as an accessory.

2. To say now what precisely is idealism, presents a considerable degree of difficulty, because of the Protean character of the object to be dealt with; but without being able to tie the wily trickster down to one shape, we may be able to effect a sufficient capture for purposes of inspection. Negatively an important observation is, that it is not idealism to maintain that the thing-in-itself is unknowable, when by thing-in-itself is meant an object out of all relation to knowledge. The stoutest realist would allow so much. But idealism has its root mainly in these two contentions, that mind cannot go outside of itself or of its own conscious states, and that least of all can mind truly represent to itself external matter. The idealist, who on these grounds should venture to affirm that there is nothing outside his thought, and especially nothing material, would be so manifestly guilty of unwarrantable dogmatism, that we may pass him by and consider only the more plausible adversary, the strength of whose position lies in its being agnostic. He does not deny, he only pleads his inability positively to affirm anything beyond the idealistic limit. This limit he may variously set according to any one of the following formulæ. I am certain (*a*) only of present states of consciousness, as of subjective coruscations or modes; (*b*) only of present along with certain remembered and certain safely expected states; (*c*) only of past, present, and future states along with my substantial mind as the subject of these

states. So far the two fundamental principles of idealism have been fairly, though in varying degrees, respected: there has been no passage beyond the thinking self, and there has been no assertion of independent matter. But many who would not dare to take up the last-mentioned of the three positions, make no hesitation in assuming the next, which is to idealism really a more formidable position, namely, (d) I am certain only of a series of conscious states which I know as my mind, and of other series which I know as states of consciousness in other minds. (e) With regard to an outer material world, some idealists, not quite thoroughgoing, claim to have a knowledge that it exists and acts upon them, but disclaim all knowledge about its real nature and properties.

The above divisions are not meant historically to represent the several schools of idealism, but rather to show progressive steps from the extremest to a more moderate doctrine. Berkeleyism, as having been already described, is omitted. In all cases idealism is founded mainly on a common difficulty which is felt against realism—a difficulty which shall now be stated in the words of an upholder of the system. The following passages, culled from Professor Caird's work on Kant, will convey the information required. "The knowledge of things must mean that the mind finds itself in them, or in some way, that the difference between them and the mind is dissolved." "How can anything come within consciousness which is essentially different from consciousness? How can we think that which is *ex hypothesi* unthinkable?" "We can know objects because in so far as their most general determinations are concerned, we produce the objects we know." Thus the one method of asserting a knowledge of things is in some way to identify thing with thought, to make thought in some way the producer of its own things, so that *esse* shall be *percipi*. If a dualism, a real division between thought and thing, is allowed, then you have thought transcending itself and reaching to something other than itself; and the only way to get over this difficulty is by some such rough-and-ready but logically unjustifiable means, as that employed by Professor Clifford, when he says that he is satisfied with his ancestry for having evolved his mode of consciousness, and adds, "How consciousness can testify to the existence of anything outside of itself, I do not pretend to declare." Thus the alternatives seem to be either to identify thing with thought, or to pass from thing to thought, as it were, by brute force; unless, indeed, we are prepared to give up the attainability of real knowledge altogether, and confess that all things are unknowable, except passing mental conditions.

3. One point, which has already been incidentally mentioned, may here be distinctly emphasized, when we are about to state what exactly we undertake to establish against idealism. In asserting that ideas cannot transcend themselves, no plausible idealist affirms that there is no transcendent reality: he only asserts the powerlessness of the mind to make

sure of it. As Mr. Bain[2] remarks in an article in *Mind*, "The statement that there is no existence beyond consciousness, is not what an idealist would make; but what he says is, that we know only what we perceive. Conscious properties make up object and subject alike: consciousness contains its object states and its subject states, and all our knowledge lies within the compass of these." In opposition to idealism as so propounded, without making special reference to the outer world of matter which was dealt with in the last chapter, we have as our substantial task to show (*a*) that there is no contradiction in the fact of the intellect, through its ideas, knowing objects really other than itself; and (*b*) that the objective reality of ideas must be admitted, because of its self-evidence, and because the fact cannot even be denied without its being at the same time implicitly asserted.

4. These being substantially the points to be made good, the requirements will be found satisfied under the following arguments and conclusions:

(*a*) Bilocation, or being present in two different places at once, is not naturally possible to a material body. This is true, but does not affect realists, who do not suppose an idea to be an extended body, which has at the same time to transfer itself to a distant space. So far, however, as an idea is indirectly subject to the conditions of space, it is

[2] On the strength of the fact that they do not dogmatically affirm that there is no reality beyond ideas, some idealists repudiate the name of idealists as applied to themselves.

physically present in only one spot, namely, in the soul united to a narrowly circumscribed body. But besides having, as all other things have, a physical entity, an idea has something else peculiar to itself, its *vis intentionalis*, as the scholastics say, its power of going forth, not mechanically, but by way of intellectual perception. Now, coolly to affirm, as idealists are in the habit of doing, that this power is unable to attain to anything outside the thinking subject, is not only the veriest piece of dogmatism, but is against the evidence of experience. Not by any *a priori* assumptions, nor by a false analogy drawn from physics, but by the accurate interpretation of conscious facts, are we to know what ideas can do. A door-post, which has no ideas, can never be taught what is the power of ideas. A man, precisely because he has ideas, can judge of their value, and his judgment must be formed on the case as presented in consciousness, not upon some hypothesis wholly arbitrary. Using the method of self-introspection, we find that our ideas are—in the wide sense of the word *things*—things having a perceptive power. Nor is there the shadow of an argument to suggest that the perceptive power cannot reach to other objects, even to objects purely material and unintelligent. As we do not know how intelligence produces its marvellous act, as that mysterious spiritual agency is above our ken, it is very arbitrary on our part to limit thought by the analogies of mechanical action. Such an attempt breaks down at every point. Even idealists themselves show the little store they set by their own theory in straightway disregarding it, and transgressing the boundaries put by themselves. Their main limitation is that thought shall not transcend itself: hence, theoretically, present consciousness, viewed as a fact, ought with them to be the whole of positive knowledge. Yet they one and all trust memory and expectation, thereby openly going beyond present fact. Few would seek escape by a hopeless attempt to deny this: hence Mill candidly confesses, "The psychological theory cannot explain memory."[3] The few, however, who are venturesome enough to make the denial, would find their bold course lead only to speedy confusion; for they would have to abide rigorously by their statement, "We know only our present conscious condition." "Very well," is the reply, "define your term 'present.' If it is an absolute, unextended point, then it is of no service to you, and is most flagrantly against the law that a certain persistence in consciousness is necessary in order to secure advertence. If your 'present' is not an unextended point, then it has a certain duration: it involves a past and a present, and you begin to be in the same condition as your bolder brethren, who openly claim to believe in memory and expectation, and who so far give up the dogma that thought cannot transcend itself."

Another surrender, and a more glaring one, of the same dogma, is the almost unanimous admission by idealists of "other men," or other consciousnesses; which is surely a full confession, that for thought to reach an object other than itself, it

[3] *Examination*, Appendix.

needs the accomplishment of no self-contradictory feat.

If considerations like the above have the salutary effect only of making the idealist less confident of his assumed position, and more respectful to the secure judgment of the *orbis terrarum;* if they only rouse him to ask himself by what right he takes it for granted, that thought must be shut up in itself, then they have been not without the beginnings of success.

(*b*) To carry these beginnings further, we may urge upon the thorough-going idealist, to whom thought is not for certain anything more than a mental firework, that he has been all along supposing the objective validity of thought in arguing out his conclusion;[4] and that his very assertion, as to the nature of ideas, is founded on the belief that his ideas concerning this point are objectively valid. On the strength of valid ideas he tries to prove ideas invalid, thus taking up the position of the universal sceptic, which we have seen to be untenable. Also we have seen that evidence is the guarantee of truth. Now to any one who will make fair use of his faculties, there is evidence for the general truth that his ideas are objectively real, even when they are about objects not actually existent, but only possible. The result cannot be the conclusion of strict demonstration, that is, of an inference from the known to the unknown. For no premisses can be framed which do not assume the conclusion. The fact, then, must be taken on

[4] Palmieri, *Logica Critica*, Thesis vi.

its own self-evidence, than which no other and no better guarantee can be given. Mediate knowledge, through means of proof, has no advantage over intuition, for it must rest finally on intuition; nor is the evidence whereby we see the sequence of an argument more valid than the evidence, whereby we assent to the simpler truths of immediate knowledge. To fancy otherwise is a common delusion with our adversaries.

But about intuition there is a confusion to be cleared up, and a mistake to be removed. Some limit intuition to the case where the object itself is actually present in the mind; as is the condition of those facts of our own consciousness, which, Malebranche says, we know "without ideas," or as the scholastics would say, through no vicarious "species." How, then, do the schoolmen, insisting on the need of the "species" for all objects outside the mind itself, yet manage to assert an intuition of some such objects? By means of the distinction, already explained, between a *signum quo* and a *signum ex quo*. Unfortunately adversaries, from a leaning to materialism, often test the case only on the merits of external bodies, about which there is admittedly a difficulty, such as to cause certain followers even of orthodox philosophy to declare themselves "cosmothetic idealists"—that is, they hold that an inference is requisite to make sure of the externality of a body. But setting aside this vexed question, we can have recourse to intuitions of truths, the objects of which are certainly not part of ourselves, and not in themselves bound up with the actual existence of an outside world of matter. Such for example are the truths contained in the propositions, "What is, cannot at the same time not be;" "Every new event must have an adequate cause." Here the ideas, "being," "not being," "event," "cause," cannot really be resolved into simpler constituents, but are seen in themselves, as soon as they are possessed, to be no idle fireworks of the mind, but to have an objective meaning, leading at once to the enunciations above made. They are *signa quibus*, a phrase fairly illustrated by some quotations to be found in Hamilton's edition of Reid's *Intellectual Powers*,[5] where, however, neither author nor editor are exactly of our mind. Take first this note of Hamilton's: "Arnauld did not allow that *perceptions* and *ideas* are really or numerically distinguished, *i.e.*, as one thing from another; nor even that they are modally distinguished, *i.e.*, as a thing from its mode. He maintained that they are really identical, and only rationally discriminated, as viewed in different relations; the indivisible mental modification being called a *perception*, by reference to the mind or thinking subject, an *idea* by reference to the *mediate* object, or thing thought." This word "mediate" should have been omitted: the immediate object of the mind, as percipient, is not primarily the idea itself—though we shall see self also entering in, when we come to describe consciousness—but it is that which is signified by the idea. This immediate object is always given intuitively, though it may

[5] **Essay ii. c. vii.**

require an inference to refer it to some larger whole, or to settle its existence in or as some actual thing. In other words, every idea has a meaning, that is, an immediate object; every idea is the intuition of an object, complete or partial. Hence Descartes is cited in the place referred to, as describing ideas to be "thoughts so far forth as they bear the character of images,"[6] and Buffier as writing: "If we confine ourselves to what is intelligible in our observations on ideas, we shall say that they are nothing but mere modifications of the mind as a thinking being. They are called *ideas* with regard to the object represented, and *perceptions* with regard to the faculty representing. It is manifest that our ideas, considered in this sense, are not more distinguished than motion is from a body moved." Besides, then, the intuitions of states of self, we may have intuitions of objects that are not self; and the view that the mind first looks at an image within itself, and then vainly tries to compare this image with some object wholly outside itself, would be very fatal to realism, if it were the true account of the process: but it happens to be a caricature, or at any rate an unintentional piece of very bad drawing.

Briefly to resume. Our refutation of idealism is, that its falsehood appears upon immediate evidence, for no one can have the normal faculties of a man without some real knowledge coming home to him, and showing him that he has really the power to know. To argue against this fact is to imply its

---

[6] *Cogitationes prout sunt tanquam imagines.*

admission. Hence, in the First Part of this book, the capability of the human mind to attain to truth was put down as the first condition to be granted at the very outset of philosophy. Ideas cannot then, as Mr. Huxley surmises, be mere flashes in the mental pan, hitting no mark, and quite ineffectual for objective knowledge. If the argument against idealism should to some appear scarcely to be an argument, the reason lies, not in the weakness of the cause, but in the fact that the case is too elementarily clear to allow of demonstration strictly so called; and in that sense alone "the opposite of idealism cannot be proved." Man, being intelligent, in the very exercise of his faculty is immediately assured of its existence and of its validity, and to ask a more roundabout proof is to demand the preposterous and the impossible. Every idea is necessarily representative or cognitive of something, and only in the rare instances, where we are reflecting upon our ideas themselves, are ideas the direct and principal objects of our intellect.

5. When we assert that the object of our ideas is real, the word "real" is very liable to misunderstanding. In a narrower sense "real" means only the actually and physically existent; but as used in this chapter, the "real" is whatever either has or might have its own physical existence, and does not exist formally as an object of thought alone, as also whatever is a real aspect of such an actual or possible entity. It is what logicians strictly understand by "a first intention," as opposed to "a second intention," that is, to an object which,

as formally described, could not exist except as the term of the mind, because the mind, with its abstractions and reflections, has imposed upon it some conditions essentially mental. Such are genera and species, subjects and predicates, and universal ideas, all which are essentially logical entities, with no more than a ground for their formation, in the extra-mental order. Besides these, everything else which is truly the object of an idea, is, in the present use of the word, "real;" though often that which is allowed to pass for an idea, is in fact no idea at all, being but a contradictory medley of ideas, never fused into one idea. It is a false judgment, or fancy, that there is such fusion between mutually repellent elements, for example "a square triangle."

6. It is useless, however, to urge the objective reality of ideas unless a special explanation is given of universal ideas, which seem to be condemned by the admitted fact, that every real object, actual or possible, is singular. Under the very false impression that all realism, when the word is used in its connexion with universals, must be of the exaggerated form, which asserts universality *a parte rei*, modern writers overlook that moderate realism which, giving to things what belongs to them, and to the mind's own operations what belongs to them, is manifestly the true doctrine.

(*a*) We shall get at the root of the solution if we observe the difference of condition between an infinite nature and a finite. The infinite nature does not allow of a multiplicity of individuals: there is but one God, and there cannot be more, for, as is shown in natural theology, a plurality of individuals, having a nature infinitely perfect, involves a contradiction, so that the three Persons are but one God.[7] But the case is altered with finite natures. Among them no one individual can claim to exhaust the possibilities of the nature; no one is so a man as to fill up, in his own person, the whole capabilities of humanity. However great the man, there is room enough in creation for others; and if "there is no necessary man," still more is there no all-exhaustive man. Any created nature, and any character about it, may be specifically repeated an indefinite number of times. In the controversy between Leibnitz and Clarke as to whether two examples of the same species can be so thoroughly alike that the only difference existing between them is that they are individually diverse, the affirmative is the right answer. Anything that has once been done may have its exact copy in another individual, yet the individualities are separate. Another Adam, in all respects *like* Adam, but not Adam, might have been the first man. But here we see reason enough why no universality *a parte rei* is possible. There *always must* be the difference that one individual is not another, while, *de facto*, besides this, there are always other differences, at least in accidentals. Nevertheless, we cling to what we have before said, and, insisting on the similarities in the midst of mentally negligeable dissimilarities, we affirm that

---

[7] This hint cannot be developed here.

the real likenesses between several creatures give the foundation for universal ideas.

(b) We have now to determine the way in which universal ideas can be formed, so as to be predicable of real things and still not to introduce any falsehood into the predication. It is certain that all the individual differences cannot be physically abstracted; such abstraction must be mental; and the mind has to be careful not to attribute its own processes to nature. By virtue of its reflective power the human intellect has a mode of coming to agreements with itself, which wonderfully serve the purposes of knowledge. Thus, being finite, it cannot directly represent to itself what an infinite object is; but by a contrivance it can obtain sufficiently an idea of the infinite; for it knows what limited being is, and it has only to deny the limit in order to form a true, though imperfect, conception of the infinite. Similarly it is by a contrivance that we fashion for ourselves a universal idea, the requisites of which are, that it shall be "univocally predicable of several individuals, taken singly or distributively." Thus "man" is predicable of Peter, Paul, John, and James: all and each are men. A direct and a reflex universal must both conspire to make up the whole. The direct universal is of "first intention:" it picks out some nature or attribute, prescinded from its individuality, as in the perfectly unindividualized conception of *virtue, vice, substance, round*. The individuality is not denied, but merely put out of the reckoning, as is indeed all "extension" of the term. Next comes the reflex universal due precisely to the addition of "extension" by the observation that a concept so prescinded may be applied to each of many individuals presenting the notes contained in the comprehension. "Mammal," let us say, is the notion we gather from the inspection of a cow; advertence to the applicability of this idea to many individuals, actual or possible, gives the reflex universal. Because of the process which forms the direct universal, the universal is sometimes called an abstract idea; and it is so inasmuch as it is always abstracted from individualizing differences. But because Pure Logic has found it convenient to define "abstract term" as one which goes a greater length in the way of abstraction, and "exhibits a form without a subject," *e.g.*, "rotundity," "humanity," "mammality;" we may respect this appropriation of a word, and say that "rotund," "human," "mammal," are *prescinded* or *abstracted* terms instead of calling them *abstract*. The abstraction, it cannot be too often remarked, is mental and not attributed to the things themselves: whereas the characters expressed by the prescinded terms are in the things themselves, and are attributed to them. It is a real predication when we say of a corpulent old gentleman that he is "human," "rotund," and "mammalian."

To go through the whole account once more in the way of illustration. Looking at a triangle, we see its essence to be a plane figure bounded by three straight lines. This is our intellectual insight into the *quiddity* or *whatness* of the thing. Any existent triangle will be scalene, or isosceles, drawn in white chalk, or in red chalk, and so forth: but content with the quiddity, we neglect these individual peculiarities, though any one of them might be singled out, and treated just as we are treating the essential triangularity itself. But to rest content with one example at a time, we have the prescinded conception, "plane figure bounded by three straight lines." This is the direct universal, universal as yet only *in potentia*, but made so *in actu*, when we recognize it, on reflexion, to be a concept which is one in many different individuals, actual or possible. There may be thousands of figures, each of which is a triangle, and admits, univocally with the rest, the predicate "triangle." The one concept, regarded as the common predicate of many, is a logical entity, a "second intention:" the direct meaning of that concept, in "comprehension," is literally true of each individual, and is "a first intention."

The whole of which doctrine is condensed by the scholastics into the phrase, "Universals are *formally* only in the mind, but *fundamentally* they are in things." Things are really like one another; and this is the foundation whereon the mind proceeds to build, when conceiving the likeness, and prescinding from individual differences, it ranks similar individuals under one common idea. We each fall under the concept "man," though no single one of us is simply "man" without individual differences, and though physically we each form no unity with other men.

(c) The objection that every idea is physically one thing, with one meaning attached to it, simple or complex, can be met by us with the reply that this holds of the direct universal, and is remedied, for purposes of universality, by the reflex act which we have described. For example, an idea of *triangle*, is one psychological state of the mind, and it has one complex signification: but on reflexion this one signification can be applied to several individuals. Hereupon we are led to remark the incompetence of the sensist theory, which accounts for universals thus: Repeated sensations from resembling bodies produce a common image by a process comparable to a recent device in photography. The photographs of several persons, for example mathematicians, are passed before the camera at such a rate, that only those features which are successively repeated by a number of pictures leave a marked impression on the sensitized plate. Other features are either lost, or but faintly indicated. The result is that a sort of average face stands out, in which enthusiasts are glad to find the resemblance of some individual who has been famous in mathematics, and who is thus proved to have had the typical countenance. By no such process could a universal idea be reached; for the average image is still singular, applicable rather to none than to all mathematicians; for even the favourite to whom it is assigned is allowed to be not accurately represented. Moreover, the photograph has no self-referring power at all: it keeps strictly at home. Assuredly there is no power in sense-images properly to abstract and universalize; and such common images as the lower animals can frame certainly do not reach to the standard of universal ideas. Hence we must insist very strongly on the strictly intellectual character of the process of universalizing, and on the fact that abstraction is no mere dropping of sensile details, without the addition of some active power of intelligence which is above sense.[8]

(d) If to form a general notion it is often necessary to multiply observations and comparisons of individuals, the reason is not that suggested by the analogy of the average photograph. One observation would suffice for the framing of any universal idea, if at once we could observe things through and through, and know all about them. One observation as to how a circle is drawn would, as a matter of fact, suffice for the universal idea of a circle, because the mode of genesis is so clear. But in physical matters we are liable to all those difficulties of generalization which are studied under the heading of Induction, and for which Mill's canons were originally devised, and have since been improved upon by later writers.

(e) The difficulties of universalizing are often so great that we do not accomplish the result, but manage to get along with terms still left in the vague. An ordinary man has never found it necessary to settle for himself precisely what he means by a tiger, a hippopotamus, or even a horse. He

---

[8] See Kant's clumsy attempt to mediate between individual sense-image and universal idea by means of his *schemata* or *monograms* of the imagination. (*Critique of Pure Reason*, Max Müller's Translation, Vol. II. pp. 124, 491.)

has vaguely outlined images of these several animals in his brain, and these suffice for ordinary purposes. If called upon to assign the precise marks which he included under each name, he would be non-plussed; the finer discrimination would be beyond his powers. A rustic, whose idea of fish was formed simply on what the hawker sold him under the pleasant name of "fresh herring," would be quite puzzled if taken into a town to see an aquarium, or even a fishmonger's shop: while a day spent with a merman "at the bottom of the deep blue sea," would utterly overwhelm him by the endless display of fishy varieties. Even a learned man may often be betrayed into calling a whale a fish, and it was a fish so far as the old usage went. In view of facts like these, we have only to say, that ideas which have never been properly abstracted and universalized must not be brought as specimens of universal ideas. There are genuine specimens, and these we must use as illustrations. We shall find them especially in mathematical and moral definitions: as also in some of those physical laws—for example, the laws of motion, which have been satisfactorily formulated.

(*f*) What has been asserted of ideas is still more applicable to words. An idea strictly is never vague: and if an idea is said to be indefinite or to vary, it is not one idea, but the addition or the subtraction of ideas, or the element of indistinctness, which is variable. Why, the mere exercise of school-boy translation was enough to teach us,

v

how far words are from having each a neatly defined signification, and the special employment of technical terms by scientific men is a contrast which calls attention to the looseness of ordinary usage. Certainly, we cannot flatter ourselves that, by the aid of a dictionary, we shall be able to read intelligently any book written in our own language, no matter how recondite the subject. Words, then, are no immediate test of the doctrine about universals.

(*g*) We may take leave of the matter with an answer to a difficulty which Mill[9] urges in this shape: In order to get your abstracted general term you must isolate its contents: but this the law of inseparable association forbids you to do: what has always been united in experience and cannot be conceived to be disunited, must always cohere in thought. Against this fancied difficulty, the power of the mind, by reflexion, to come to agreements with itself, must once more be insisted upon. To abstract a common nature or a common attribute, it is not necessary to shut out concomitant ideas of individual peculiarities; it is quite enough to know which are the common notes, and to resolve to take account of them alone. It is possible in society to ignore the presence of a man, of which yet you are aware. If any one has the general notion of a plane triangle as a plane figure bounded by three straight lines, it in no way stops his reasonings upon this abstracted nature, if there is concomitantly in his imagination, or in his thoughts, the

[9] *Examination*, c. xvii. pp. 320, 321. Contrast St. Thomas, 1a. q. 85, a. 2, ad 2am.

representation of scalene or isosceles properties. These may be present to the mind and yet wholly left out of count in a selected line of thought. Otherwise all reasoning would be baffled: for we always have an accompaniment of variously suggested ideas going along with the main ideas, but excluded from entrance upon the course of argument. Whatever may be our doctrine about the number of thoughts that can be present to the mind at one time, we must find room for that familiar experience, whereby consciousness has its point of greatest attention surrounded by a region of diminishing advertence, and shades off into the subconscious and the unconscious. There is one brightest spot, and round it there is a fainter halo: there is a substantial vesture of thought, and to it adheres a fringe. But we can abstract what part of the whole we like, by our will to do it. Ideas need not be in our mind like so many sharply distinct atoms: they may be there after the analogy of parts in a network or in an organized body, and yet we can fix upon such a portion as we choose, and equivalently isolate it. Mill himself allows that we can so do, though he makes a great fuss about the inseparability of uniformly associated ideas:[10] "The formation of a concept does not consist in separating the attributes, which are said to compose it, from all other attributes of the same object. We neither conceive them, nor think them, nor cognize them in any way, as a thing apart, but solely as forming, in combi-

[10] *Examination*, l. c.

nation with other attributes, the idea of an individual object. But though thinking them only as part of a larger agglomeration, we have the power of fixing our attention on them, to the neglect of the other attributes with which we think them combined. While the concentration of attention actually lasts, if it is sufficiently intense, we may be temporarily unconscious of any other attributes, and may really, for a brief interval, have nothing present to our mind but the attributes constituent of the concept. In general, however, the attention is not so completely exclusive as this: it leaves room in consciousness for other elements of the concrete idea. General concepts, therefore, we have properly speaking, none; but we are able to attend exclusively to certain parts of the concrete idea, and by that exclusive attention we allow those parts to determine exclusively the course of thoughts as called up by association." If Mill would only cease to make mind so much of a mere machine, and if he would make it, instead, an intellectual faculty proceeding on insight, with a vast power of spontaneity, with a power to reflect, to abstract, and to come to agreements about its own operations: and, if further he would observe that to think certain characters apart need not mean, and does not mean, the same thing as to think that in real objects these characters do actually exist apart; then he would have little scruple in revoking that portion of his own declaration: "General concepts we have properly speaking none." Also he would make less of the necessity for an association with words, such that "the association of the particular set of attributes with a given word is what keeps them together in the mind, by a stronger tie than that with which they are associated with the remainder of the concrete image." If only he could have formed a truer conception of how human intelligence works, and had taken warning in season from the necessity under which he found himself to make such confessions as, "I have never pretended to account by association for the idea of time," Mill would have ceased to regard it as a misfortune, that mankind ever took up the expression, "General conception."

7. The object of this whole chapter has been to defend the objective validity of ideas in general; but not of course to say, in detail, what ideas in each science are the correct representatives of reality. The main root of difference between adversaries and ourselves, is that they will insist, contrary to us, in regarding knowledge as primarily not a knowledge of things but of ideas. They imagine that what we first of all know are always subjective affections as such—*signa ex quibus* and not *signa quibus*—and then of course they see no way to a proof that these subjective affections are like objects without; rather they are inclined to believe that there can be no likeness, but at most a symbolic correspondence. But this is not the legitimate interpretation of the doctrine that the mind perceives through ideas. The mind perceives through ideas, not in the sense that it looks at ideas first, and then passes on to infer things; but in the

sense that the mind, at least under one aspect, begins as a *tabula rasa,* and only in proportion as it stores itself with ideas is it rendered by them cognisant of objects. The mind, as informed by an idea, is cognisant of an object: but the idea as has been so often repeated, is a *signum quo,* not *signum ex quo;* it has not first to be known, but is itself constitutive of the act of knowledge. A world of misconceptions would be saved if the right view of the office of ideas were acquired—misconceptions which have led to the false definitions of truth exemplified in our opening chapter. In support of our own definition we need only a right appreciation about the nature of ideas; then ideas are seen to be objectively valid, and true knowledge is perceived to be the conformity of thought to thing. We thus escape the deduction from Helmholtz's theory of sensation—the deduction, namely, that our sensations being non-resembling signs of external things, all our ideas are non-resembling signs so far as they concern objects outside ourselves. Briefly, we recognize that we have a power of real knowledge, not reducible to a mechanical reaction, or *quasi*-chemical combination.

### Addenda.

(1) It is a fancy of some semi-idealists that the thing-in-itself is something out of all relation to knowledge, and therefore not knowable for what it is. The mind gives to this unintelligible thing a form of its own, frames a symbol for it, but symbol and symbolized have nothing alike between them.

(2) The supposed impossibility of *knowledge* trans-

cending the conscious state is really not kept to, by those who profess to keep within the impossible limit. Thus Mr. Spencer [1] has to have recourse to all the convenience of *knowledge* extending beyond the conscious state, under the subterfuge of calling this knowledge by another name. He says, "though *consciousness* of an existence, which is beyond consciousness, is inexpugnable, the extra-conscious not only remains *inconceivable* in nature, but the nature of its connexion with consciousness cannot be truly *conceived*. Ever restrained within its limits, but ever trying to exceed them, consciousness cannot but use the forms of its activity in figuring to itself that which cannot be brought within these forms." Thus we are *conscious* of an outer reality which we do not *conceive* or *know*. The artifice here is ingenious but unsatisfactory; any fact which *consciousness* enables us with certainty to predicate, deserves to be called *knowledge*.

(3). The word "intuition" has been employed above with a risk of misinterpretation. For, not to mention other views, on a theory given more or less explicitly by different writers, an "intuition" stands for an implanted instinct to believe something, without either immediate or mediate evidence. As used in this work, an intuition is no innate idea or perception, and no specially communicated knowledge: it is simply knowledge on immediate evidence. An instance in point is man's perception that his ideas have objective validity; on perceiving a clear truth he has an intuition of the validity of his faculties; and without this intuition he never could ascertain the fact by strict process of inference.[2] There are, moreover,

[1] See the opening chapters of *First Principles*.
[2] Recur to what is said in the body of this chapter about intuition (pp. 313—319).

intuitive perceptions beyond this matter of self-consciousness and in the region of the *non-ego*.

(4) Another point, already touched upon, may be further elucidated. With logicians an "abstract." idea is strictly one representing "a form without any subject," *e.g.*, "humanity." But any universal term, abstracted from individual peculiarities, is often called abstract, *e.g.*, "man." The fact is there are degrees of abstraction increasing in extent: from the concrete article in his hand the bowler, at a cricket-match, may progressively abstract the terms "ball," "spherical," "sphericity." Only the last of these words is an abstract in the full sense required by Pure Logic. With Hegel any word, not significant of the whole universe, was an abstract term, so complete did he make the unity of the whole. Thus, as we are not omniscient, all our knowledge would be abstract.

(5) In admitting that the mental process departs, in the formation of universal ideas, from strict reality, we are only allowing the mind to do what it often does without risk of falsehood. In nature the line of progress is from causes to effects: in our knowledge the progress ordinarily is from effects to causes; what logically is the premiss to a conclusion is often, in the ontological order, a consequence of the fact, or the principle, stated in the conclusion. We may argue God's wisdom from the order in creation, but the order in creation is a consequent upon the Divine wisdom. Again, we often make mental distinctions where we know there is no real distinction: as when we divide God into a nature with distinct attributes. Any departure, therefore, which in the formation of universals is made away from reality, can be recognized as such, and need not be asserted of the reality. To the real that alone need be assigned which belongs to it.

(6) Hence we know what to reply to those who, like Professor Huxley, maintain that our generalized laws of nature are not real but ideal. It is true that, supposing the law to be correctly formulated, there is no general law of gravitation apart from the several particles of matter which attract; but as each and all do attract, the universalized law is real in all that it attributes to nature. The difficulty is solved in the general solution of the problem concerning the reality of universal ideas; and to declare that generalized laws are not real, is a statement more likely to mislead than to instruct. They are real so far as they are applied to nature, and have their foundation there.

(7) Now that we are coming to an end of the doctrine about universals, we may observe that there seems more difficulty about individualizing our ideas than about universalizing them. The Divine nature excepted, every other term, in its mere statement, might belong to an indefinite number of individuals. "The first man" might have been quite another; and all that we have recorded of Julius Cæsar might have been verified of another man, down to the minutest detail, which human description can record. For we never have an intuition of individuality itself as such. Our demonstrative pronoun itself, backed up by additional terms, "this very individual," is left a universal, unless we can fix it, proximately or remotely, by some fact of concrete experience. Touch a thing, while you call it "this," and you are fastening upon an individual; but mere ideas without an experienced connexion in fact,—either your own experience or the experience of some one else, —will not carry you out of the universal. "This man" has no individuality till it is somehow concreted in experience.

(8) The true doctrine about realism was settled very

early in the course of the scholastic disputations; not that some did not continue to go wrong, but the right statement was elicited and widely recognized. This is a point on which it is hopeless to consult an ordinary non-scholastic author; as soon as ever you see him starting the subject of the old controversy about universals, as a rule you may say to yourself, " Now for some quite incompetent criticism, and a large display of ignorance." As a single specimen of one who early formulated the doctrine of moderate realism we will take neither Albert the Great nor St. Thomas, but a contemporary Dominican, the preceptor in the home of St. Louis of France, Vincent of Beauvais.³ " Universals," he writes, " are not in the intellect alone. For men have one common undivided nature, which is humanity, by reason of which each is called man; and that which is thus participated by all is called universal." Realism of the most extravagant type! the reader will perhaps exclaim; but let him have the patience to continue. " What is common is their *specific* likeness, which *by the intellect is taken in abstraction from the individualities.* For as a line cannot exist apart from matter, and yet the intellect makes no false judgment when it abstracts the line from the matter, because it does not think that the two are really separable, but merely thinks of the line without taking account of the matter; so in general any universal, though it cannot be apart from its singulars, yet can become an object of intelligence, while no attention is being paid to what is individual." This clear explanation invites comparison with modern

[3] Quoted by Stöckl, *Geschichte der Philosophie*, under the name *Vincent de Beauvais*. Compare how this doctrine differs from Mill's popular fallacy about the scholastic doctrine. (*Logic*, Bk. I. c. vi. § 2.) He has the effrontery to put down exaggerated realism as "the most prevalent philosophical doctrine of the middle ages." (*Examination*, c. xvii. pp. 308, 309.)

statements, such as that of Dr. Maudsley, when he says, that while "no animal, as far as we can judge, is capable of forming an *abstract* idea, there is good reason to think that the more intelligent animals are able to form a few *general* ideas." Generalization without abstraction is impossible, if the author is speaking strictly of a general idea. To return to the mediævalists, however; they so talk of abstracting the essence from the individual accidents, that a reader might suppose they confined universals to essential predications. But though they thus emphasize one of the most important cases of universalization, they fully allow that any attribute may be abstracted and made a universal term; but in all instances alike this will be considered in its *quiddity* or nature, for accidents also have their *quiddity*.

(9) The objective validity of ideas once established, it is not necessary explicitly to argue that judgments and reasonings are valid processes, when they properly embody these ideas. Distinct propositions on these subjects may be found in the ordinary text-books;[4] but it is not difficult for any intelligent reader to guess the substance of the arguments employed.

[4] Palmieri, *Logica Critica*, Theses xiv., xviii.

# CHAPTER IV.

### EXAGGERATED REALISM, NOMINALISM, AND CONCEPTUALISM.

*Synopsis.*
1. Exaggerated realism.
2. Nominalism. (*a*) Nominalists assert that universality is only in the word, but do not deny real likenesses between things. (*b*) Refutation of nominalism. (*c*) Specimens of nominalists in England.
3. Conceptualism. (*a*) How conceptualists improve upon nominalists. (*b*) Refutation of conceptualism.

1. THE error, which is often confounded with the realism defended in the last chapter, is the doctrine of exaggerated realism. Any theory which asserts a formal universality *a parte rei*, which supposes, for example, that there is some concrete nature physically common to all men, and only accidentally individuated in each, must be rejected as wanting even in intelligibility. Such cases as that of substance permanent under its varying activities and passivities, are in vain quoted as examples of universality *a parte rei*: while Cousin's assertion, that space is a real universal, shows him to have entertained crooked notions either about space or about universals. No pretended instance can stand testing: and if some mediæval philosophers thought

otherwise, we give them up and say, they were mistaken; but it hardly becomes certain modern critics to make merry at the expense of the middle ages, when they themselves are in favour of *monism*, a single underlying reality, of which all that we experience, and we ourselves, are but the phenomena.

2. (*a*) In extreme opposition to the exaggerated realist is the nominalist, who, if thorough-going, places universality in the name only. Not that nominalists deny a real likeness between things, for that is too obvious to be gainsaid; and Mill finds fault with Hamilton, whom he supposes to hold that such likeness is denied. Hobbes, a notorious nominalist, says clearly enough, that "one universal name is imposed on many things *for the similitude in some quality or other accident.*" Indeed, the perception of similarities and dissimilarities is made by some nominalists to be the very basis of all knowledge.

(*b*) The state of the case is, then, that while admitting real similitudes and our knowledge of them, nominalists have so far ignored these in their account of universals as to declare, that the universality is only in the word, and neither in the things nor in the concepts. That it is not *formally* in the thing we admit; that there is *no foundation* in the thing we deny, for there is the real likeness, affording to a mind which has the power of abstraction and reflexion, a *groundwork* for the formation of universal concepts. Next we affirm that the universal formally, or as such, is in the concept, or in the arrangement of concepts already described, as respectively direct and reflex universals. If it were not there, it could never be in the word; or if it were in the word, and not in the concept, it would never enter into knowledge. Besides, it is absurd to suppose a word, as such, to be universal: for the spoken sound and the written character are conventional signs, and always in themselves singular, no matter how often repeated. Each repetition is individual: only the mind can universalize a sign, and its power so to do is evident from our previous explanation of the process.

(c) These facts are so obvious that it becomes necessary to give evidence that there are professed nominalists who, whatever their consistency, do promulgate the doctrine here refuted. "The universal," says Hobbes,[1] "is neither something existing in nature nor an idea, nor a phantasm, but always a name." Berkeley[2] sets out from nominalistic principles: "As it is impossible for me to see or feel anything without an actual sensation of that thing, so it is impossible for me to conceive in my thoughts any sensible thing or object distinct from the sensation or perception of it." He disavows the power of abstraction, without which thoughts cannot be universalized: "Whether others have this wonderful faculty of abstracting ideas, they best can tell;" as for himself he can variously compound individual parts, but cannot rise above the individual. Mixing up sensitive imagination, which of

[1] His doctrine may be found, *De Corpore*, c. ii.; *Leviathan*, Part I. c. iv.
[2] *Principles of Human Knowledge*, Introduction, § 15, 16.

course cannot duly perform the office of abstraction, with thought proper, he says : " For myself I find, indeed, that I have a faculty of imagining, or representing to myself the ideas of those particular things I have perceived, and of variously compounding and dividing them. I can imagine a man with two heads, or the upper parts of man joined to the body of a horse. I can consider the hand, the eye, the nose, each by itself, abstracted or separated from the rest of the body. But, then, whatever hand or eye I imagine, it must have some particular shape or colour. The idea of man that I frame to myself, must be either of a white, or a black, or a tawny, a straight or a crooked, a tall or a low or a middle-sized man. I cannot by any effort of thought conceive the abstract idea of man, motion," &c. All this talk is an utter ignoring of the power of reflective thought to pick out what it chooses, to fix upon a definition, and to deal with that as with a mentally isolated part. Hume continues the tradition taken up from Berkeley, whose doctrine on universals he pronounces[3] "one of the greatest and most valuable discoveries made of late years" —a discovery which he himself seeks to "put beyond all doubt." He frames the theory thus: "All general ideas are nothing but particular ones, annexed to a certain term, which gives them a more extensive signification, and makes them recall, upon occasion, other individuals which are similar to them. A particular idea becomes general by being annexed to a general term, that is, a term

[3] *Treatise on Human Nature*, Bk. I. Part I. sec. vii.

which, from a customary conjunction, has a relation to many other particular ideas, and generally recalls them in the imagination. Abstract ideas are, therefore, in themselves individual, however they may become general in their representation. The image in our mind is only that of a particular object, though the application of it in our reasoning is the same as if it was universal." This is inadequate and wrong *de more*. Mill,[4] of course, follows in the wake of Hume, and we have already heard him declare: "General concepts we have properly speaking none: we have only complex ideas of objects in the concrete;" and by exclusive attention to parts of an associated whole, "we can carry on a meditation relating to the parts only, as if we were able to conceive them separately from the rest." This power of separate conception, so far as we approach to it, he attributes to the association of the separated characters with a word, instead of to the mind's power of abstraction, or *præcisio objectiva*.

3. (*a*) Conceptualists allow that the universal is in the idea, but deny its objective reality. They can gainsay the real likenesses between things no more than can the nominalists; but they do not perceive that herein is a *foundation* for all the objective reality which we want. Where they improve on the nominalists is in admitting the possibility of a universal idea; a result which comes from their having a better theory of mental action. This improvement is strongly to be accentuated,

[4] *Examination*, c. xvii. p. 321.

and shows the large step from nominalism to conceptualism. Mental action, according to the nominalists of this country, is tied down to sensations, and to mechanical or chemical associations of ideas. Instead of a voluntary power of abstraction, they assert a "law of obliviscence," a loss to consciousness of one part of a complex aggregate, through an excessive attention to another part. As in the matter of human will they allow only a conflict and final preponderance between concurring attractions or repulsions, while we assert an intellectual power to consider the *pros* and the *cons* of separate courses, and a power of free choice supervening: so in the matter of general ideas they ignore, while we and conceptualists insist upon, the spontaneous activity of the mind in taking up, or leaving alone, elements in an aggregate conception, according to the purpose in view. Thus conceptualists are enabled to abstract from individualizing differences and to universalize what they so acquire. In spite of their better premisses conceptualists arrive at a false conclusion: but it is something that they excel the nominalists by admitting universality in ideas, while their mistake seems often a mere oversight rather than a rooted error.

(b) Conceptualism is wrong in that it pushes a truth too far: it sees that there is no formal universality *a parte rei*, and thereupon it sweepingly denies the objectivity of universal ideas. A distinction is needed. The universal ideas in what they represent are objectively real; but not in their abstract mode of representation, which, however, is not predicated of objects. When of any individual it is predicated that he is a man, the predicate is, we will suppose, strictly applicable: but no individual is man in the abstract. As, however, the individuality is not pointed out by the universal term, so neither, on the other hand, is the individuality denied: it is simply omitted. The case is made all the clearer by the reality of the physical sciences. When we are told that the best scientific generalizations are not real, we reply that this is going too far. Supposing them properly made, they are real, in the sense in which, against conceptualism, moderate realism is true. Any one who has appropriated to himself the correct doctrine of universals, has got the means of exactly determining how far a legitimately generalized law is a real law. The laws of motion, for instance, represent a part of the reality of nature, even though they be not, perhaps, three distinct laws, but only a threefold enunciation of results, due to one common principle and even though their enunciation by us be incomplete as a statement of the whole case. Perhaps there is some simple law of action at work in nature, which law, if comprehended, would give us all that we know under our three laws of motion and a good deal more besides. Still, as partial solutions of a complex problem, the three laws are really true: for they sum up experienced facts, and they do not necessarily involve anything not in the experience. Even if we make our simple starting-points what are really not primal elements but resultants from compound forces, still, as we never declare our ultimates to be absolutely ultimate, but only ultimate for us, we keep on safe ground. So some suppose that the law of attraction, as formulated by us, may be not elementary but a resultant; be it so, and it remains a real law—a law of derivatives, if not of primitives. A being who could ascertain the attraction of a large spherical planet only as something proceeding as if from the centre, not as really due to every single particle, would be right as far as he went. In such a way do we maintain the reality of generalized laws in physics. A scientific man, in his own interest, should be slow to clutch at a theory either of nominalism or of conceptualism; whereas he may be quite happy if he can intellectually justify to himself moderate realism.

# CHAPTER V.

## CONSCIOUSNESS.

*Synopsis.*
1. Some differences of definition.
2. Some differences of doctrine, especially on the question, Are there any unconscious thoughts? (*a*) Authors, really or apparently, on the affirmative side. (*b*) Authors, really or apparently, on the negative side.
3. Some settlements on the subject of consciousness. (*a*) The meaning of self. (*b*) Consciousness is found improperly in the sensitive order, properly in the intellectual; the two orders must be carefully distinguished. (*c*) The connexion between the two orders, when a man becomes intellectually conscious of his own sensitive states. (*d*) Enumeration of the objects of consciousness, and defence of the validity of consciousness in regard to its objects.

*Addenda.*

1. AT the outset many differences of definition, accompanied by some real divergences of doctrine, perplex the inquiry into consciousness. We will begin with the matter of definition, not so much seeking to exhaust the list of actually proposed definitions, as to show a possible scale of increasing contents in the meaning assignable to the term defined. First, consciousness may be made to signify no object beyond the simple fact that we are aware of our own thoughts and feelings as they occur. Next, we may include in consciousness,

besides the states just mentioned, the substantial subject of which they are the modifications, and which upon reflexion is manifested, not indeed in itself alone, but in these very affections or activities of its own. Thus consciousness would embrace the substantial self and its immediately perceptible states while these latter lasted. A trust in memory and expectation carries consciousness still further beyond present states of self to past and future. Fourthly, we may widen consciousness to the compass of all known objects, whether self or not-self, provided such objects be present at the time to the faculties;[1] so that, in the language of Hamilton, we should be *conscious* of last week's concert only as an image retained in the memory, but for the reality of the past fact we should have to depend on *belief*. Lastly, we may abolish this distinction between present and non-present, and declare that whatsoever object we know, of that we are conscious.[2] Distinguishing between consciousness and self-consciousness, some prefer to say, that while we are merely *conscious* of any outer object which we happen to know, we are *self-conscious* of a headache, a mental anxiety, or any other internal

[1] Hamilton says, "Consciousness and immediate knowledge are universally convertible terms: so that if there be an immediate knowledge of things external, there is consequently the consciousness of an outer world." (*Discussions*, p. 51.)

[2] Hamilton makes some accommodation even for this wide usage: "Consciousness comprehends every cognitive act: in other words, whatever we are not conscious of, that we do not know. But consciousness is an immediate cognition. Therefore all our mediate cognitions are contained in our immediate." (*Reid's Works*, p. 810.)

state of our own. So far for matters of choice in the definition of a term.

2. We must now approach real disputes, and begin with that about the existence of unconscious intelligence. Some would make the test of the presence of reason in any substance its power of adjusting itself to ends; in such sort that a growing plant, and a developing animal germ would be said to reason out their evolution. Kant, while he will not say that organic processes are intelligent, would have us look upon them in that light as an aid to our understanding, when we consider the operations of living matter; and the same artifice he would extend to the workings of merely physical law. Many others also show the like tendency to attribute some dark kind of intelligence to the self-arranging powers of matter in chemical reaction; and this tendency is specially natural in those who regard the elements of matter as primitive "mind-stuff," only needing a certain degree of organization to cause it to wake up into consciousness. Long ago Telesius and Campanella—and they were not the first—supposed an obscure knowledge to reside in minerals and plants. Each of the monads of Leibnitz was supposed to reflect within itself all the universe; the difference being, that some monads were as in a deep sleep, others as in a dream, others as in full wakefulness. Many evolutionists, however, attribute no cognitive power to natural objects till something higher than the lowest ranges of the animal kingdom is reached; and even here they would regard the mere organic processes as not

cognitive. We are thus brought across a question which is far more than a matter of the definition of terms; the conflict is between two most opposite doctrines as to the source of intellect and consciousness—whether consciousness springs directly from unconscious intelligence, and remotely from the non-intelligent.

The dispute however which specially concerns us turns on the point, whether there can be sensation, thought, and volition without consciousness. Those who answer in the affirmative, occasionally make of consciousness a distinct faculty; but now-a-days they would more generally be content with maintaining that consciousness depends on the relative degree, or intensity, of the act of which we are said to be conscious. It will be instructive to listen to a few testimonies on both sides; on the part of those who affirm, or seem to affirm, and on the part of those who deny, or seem to deny, that consciousness is bound up with every sensation, thought, and volition.

(a) Hutcheson,[3] who is accused by Hamilton of making consciousness a distinct faculty, at least teaches that all sensations and thoughts are conscious; and Reid,[4] who does indeed make consciousness a distinct faculty, or "different power," when he is describing it, gives no hint that he admits such a thing as unconscious thought, and in the second of the given references he says expressly,

---
[3] See *Hamilton's Reid*, note H, p. 929.
[4] Reid, *Intellectual Powers*, Essay i. c. i.; Essay ii. c. xiii. p. 223.

"consciousness always goes along with perception."[5] In Leibnitz, however, we have an author, who, besides speaking of unconscious ideas in minerals and plants, held that in man there were unconscious perceptions, or, as he expresses it, perceptions without apperceptions. Ferrier is very insistent, and rather mystic, in the way in which he distinguishes consciousness from sensation and reason. So far as a single passage can be illustrative, perhaps, the following is one of the best; but it must be remembered that the sense attributed to the word consciousness is peculiar:[6] "What do we mean precisely by the word consciousness, and upon what ground do we refuse to attribute consciousness to the animal creation? In the first place, by consciousness we mean the notion of self; that notion of self, and that self-reference, which in man generally, though by no means invariably, accompanies his sensations, passions, emotions, play of reason, or states of mind whatsoever. . . . The presence of reason by no means necessarily implies

---

[5] Hamilton (*Lectures on Metaphysics*, Vol. I. p. 212) says: "Reid and Stewart maintain that I can know *that* I know, without knowing *what* I know." Two faculties may be distinct, yet always act together. But this is not the assertion of unconscious thought: and it is precisely the complaint of Hamilton against Reid and Stewart, that the latter "had not studied, he even treats it as inconceivable, the Leibnitzian doctrine of what has not been well denominated *obscure perceptions* or *ideas*—that is, acts and affections of the mind, which manifesting their existence in their effects are themselves out of consciousness or apperception." (*Reid's Works*, p. 551.)

[6] *Introduction to the Philosophy of Consciousness*, Part I. c. v. pp. 39, 40; *Institutes of Metaphysics*, passim.

a cognisance of reason in the creatures manifesting it. Man might easily have been endowed with reason, without at the same time becoming aware of his endowment, or blending with it the notion of himself." The context shows that reason is not here employed in its ordinary sense; we had better pass on to the plainer terms of Mr. Bain, who says:[7] "Consciousness is inseparable from feeling (*i.e.*, Sensation and Emotion), but not, as it appears to me, from action and thought. True, our actions and thoughts are usually conscious, that is, known to us by an inward perception; but the consciousness of an act is manifestly not the act, and, though the assertion is less obvious, I believe that consciousness of a thought is distinct from the thought. The three terms, Feeling, Emotion, and Consciousness, will, I think, be found in reality to express one and the same attribute of mind . . . which is the foremost and most unmistakeable attribute of mind." Thus knowledge and feeling are distinguished, and the latter, not the former, is made the essential fundamental act of mind; on which theory we may conceive a mind blindly feeling without knowledge of an object, yet conscious of the feeling. Lewes holds that, "we often think as unconsciously as we breathe." His theory of consciousness is thus stated:[8] "Consciousness and unconsciousness are correlatives, both belonging to the sphere of sentience.

[7] *The Senses and the Intellect*, c. i. in initio. The doctrine is repeated in *Mental Science*, note E.

[8] It will be enough to read the fourth chapter in Problem iii. in *The Physical Basis of Mind*.

Every one of the unconscious processes is operant, changes the general state of the organism, and is capable of at once issuing in a *discriminated* sensation, when the forces which balance it are disturbed. I was unconscious of the scratch of my pen in writing the last sentence, but I am distinctly conscious of every scratch in writing this one. Then as now, the scratching sound sent a faint thrill through my organism, but its *relative intensity* was too faint for discrimination; now that I have redistributed the co-operant forces, by what is called an act of attention, I hear distinctly every sound the pen produces. The consciousness—by Descartes erected into an essential condition of thought—was by Leibnitz reduced to an accompaniment, which not only may be absent, but in the majority of cases is absent. The teaching of most modern psychologists is, that consciousness forms but a small item in the total of psychical processes;" a doctrine illustrated by George Eliot in the important part which that author makes unconscious influences exert in the play and the formation of character. Turning to Dr. Maudsley,[9] we find the following confirmatory sentences: "It is a truth which cannot be too distinctly remembered, that consciousness is not coextensive with mind, but is an incidental accompaniment of mind." And again, "It seems to me that man might be as good a reasoning machine without as with consciousness. It is only with a certain intensity of representation,

---

[9] *The Physiology and Pathology of the Mind*, c. i. p. 15. (Second Edition.)

or of conception, that consciousness appears." Such opinions are largely prompted by pathological cases, in which the patients go through their routine actions as if they were unconsciously rational. A French soldier, wounded in the Franco-German War, has furnished a very striking example. Mr. Huxley allows that possibly he is conscious, in spite of appearances to the contrary. In a different category, yet bearing on the same opinions, and illustrating the old law that objective perception, and subjective advertence to self, are in inverse proportion, stand some words of Cardinal Newman, which shall close the quotations on this side of the controversy.[10] "In what may be called the mechanical operations of our minds, propositions pass before us and receive our assent without our consciousness. Indeed I may fairly say, that those assents, which we give with a direct knowledge of what we are doing, are few compared with the multitude of like acts which pass through our minds in long succession, without our observing them. That mode of assent, which includes this unconscious exercise, I may call *simple assent;* but such assents as must be made consciously and deliberately, I call *complex* or *reflex assents.*" Scientific certitude is thus "the perception of a truth with the perception that it is true, or the consciousness of knowing as expressed in the phrase, I know that I know."

(*b*) If we omit the discussion of mere sensitive action, and confine ourselves to the main point,

[10] *Grammar of Assent*, Part II. c. vi.

intellectual action strictly so called, it is certainly the doctrine of St. Thomas, that all thought must be consciously referred to self, though the advertence need not be very explicit. That such is his teaching may be gathered from what has already been explained in Part I. chapter ii. concerning his doctrine about judgment, namely, that when the mind judges, it implicitly affirms the consciousness of its own knowledge. And a more general assertion of the inseparability between thought and consciousness may be found in the *Summa*.[11] At the same time it is well to remember, that the disputes about consciousness as a special element, or aspect, in mental life, belongs rather to recent times.

It may be well to cite here one or two English writers on philosophy who proclaim that consciousness must ever go with thought. Locke,[12] in the course of his well known contention, that we could not have innate ideas without being aware of them, writes: "It is altogether as intelligible to say that a body is extended without parts, as that anything thinks without being conscious of it, or perceiving that it does so." Dr. Brown[13] may be quoted for the same opinion, though his main effort is bent on the proof of what is not quite the same thing, namely, that consciousness is not a distinct act or faculty. If Hamilton is put in the same class, it must be with the reservation that what he says about latent thought, and about the difference between knowledge and the blind

[11] Part I. quæst. lxxxvii. art. i. et iii.   [12] Bk. II. c. i.
[13] Lecture xi. at the end. Cf. Stewart's *Elements*, Part I. c. ii.

element, belief, considerably takes off from his value as a witness. For example, he teaches that "to know is to know that we know," and in note H, already referred to, he lays it down, that "while knowledge, feeling, and desire, in all their various modifications, can only exist as the knowledge, feeling, and desire of some determined subject, and as this subject can only know, feel, and desire inasmuch as it is conscious that it knows, feels, and desires, it is therefore manifest that all the actions and passions of the intellectual self involve consciousness as their generic and essential quality." On the other hand,[14] he declares his firm conviction that there are unconscious "mental activities and passivities;" but then he seems careful not to call these "thoughts" or "cognitions," but only "modifications" of the mind; which modifications if they were referred only to material processes in the brain, helpful to thought, and were literally "unconscious cerebrations," could be allowed without demur. A more uncompromising witness than Hamilton is found in Dr. M'Cosh: "I believe that we are momentarily conscious of every sensation, idea, thought, or emotion of the mind." Any appearances to the contrary he attributes to faintness of advertence and lapse of memory.

In using authors of the school of pure empiricism,

[14] *Metaphysics,* Lecture xviii. Mill declares that there "is no ground for believing that the *Ego* is an original presentation of consciousness" (*Examination,* c. xiii. in initio), and in the Appendix to Reid, p. 932, Hamilton says: "Consciousness is, first, the mental modes or movements themselves *rising above a certain degree of intensity.*"

we must remember the deductions to be made for men who cast doubt on the substantial self, and assert only series of states unaccountably linked together in consciousness. Under these drawbacks the two Mills[15] may be quoted; the father as saying, "To feel a sensation is the sensation, to be conscious of an idea is that idea;" and the son as praising his father's words. From the same school we have also Mr. Huxley teaching that "there is only a verbal distinction between having a sensation, and knowing that one has it."

If now we may leave our insular for continental writers, we have an example in Spinoza,[16] who says: "As soon as any one knows a thing, by that very fact he knows that he knows, and knows simultaneously that he is conscious that he knows what he knows, and so on *ad infinitum*." Kant[17] declares that no object can be perceived or conceived, unless through the unity of consciousness; and adds: "It is the *one consciousness* which unites the manifold which has been perceived successively. This consciousness may often be very faint, and we may connect it in the effect only, and not in the act itself, with the production of a concept. But in spite of this that consciousness, though deficient in pointed clearness, must always be there, and without it concepts, and therefore knowledge of objects, are

---

[15] *Examination*, c. viii. p. 115.
[16] "Simul ac quis aliquid scit, eo ipso scit se scire, et simul scit se scire quod scit, et sic in infinitum." (*Ethics*, Part II. Prop. xxi. Schol.)
[17] *Critique of Pure Reason* (Max Müller's translation), pp. 92—97, 277, 278.

perfectly impossible." Cousin, in the lectures already quoted, though in one place he professes to leave the question open, yet speaks as if his impression were, that all thought must be conscious: "It is the fundamental attribute of thought to have consciousness of itself. Consciousness is the inner light which illumines everything in the soul—the accompaniment and the echo of all our faculties." And in his *Introduction à L'Histoire de la Philosophie*,[18] he teaches that "intelligence without consciousness is the mere abstract possibility of intelligence, not actual intelligence."

3. Enough has now been adduced to put the reader in a position for seeing how the dispute lies in the controversy about the nature of consciousness as distinguished from other terms. Probably the divergence between some of the writers, who have been ranged on the opposite sides, is not as great as might at first sight appear. But it is time to be laying down our statement of the true doctrine: for we must so far explain and defend consciousness as to warrant, in general, its use for the acquisition of certitude.

(*a*) Consciousness signifies the reference of some mental state to self: and what precisely we mean by self has first to be settled. A thorough-going idealist,[19] who confines himself to ideas as succes-

---

[18] Leçon 5me, p. 97.

[19] Mr. Bain's attempt to distinguish "object consciousness" as "putting forth energy," and "subject consciousness" as "pleasure, pain, and memory," is not very happy. (*Mental Science*, note E.) He says man's body belongs to the object world.

sive phenomena, ought to call self the subjective aspect of these ideas, and not-self the objective aspect; considered as so many acts of thinking, the ideas form the self, while these same ideas viewed on their reverse, or objective, sides, would constitute the not-self. In the phrase, "My thoughts about things," "my thoughts" would be self, "about things" would be not-self. At least this is the only consistent course for idealism pure and simple, which is at the same time phenomenalism pure and simple. There is here no substantial soul and no substantial body included under self. A system a degree better would admit, within the self, a substantial principle, either a spirit only, or a compound of matter and spirit. Lastly, the true meaning of self, which is vindicated partly in various passages of this treatise,[20] and partly in the treatises on General Metaphysics and on Psychology, is the composite substantial man, immediately aware of a number of bodily and mental phenomena as belonging to himself, and aware of his continuous personal identity. Though it clearly requires reflexion to bring out the element by analysis, man is immediately conscious of his own substantial *Ego*, not in its unmodified condition, but under its perceptible modifications: and what is called the "logical unity of consciousness" gives, notwithstanding Kant's denial, the fullest warrant for assuming "the substantial unity" of the thinking subject.

(*b*) If, therefore, we regard the self as a com-

[20] See, for example, Bk. II. c. i.; Bk. I. c. xi. Addenda (1).

pound of body and soul, in examining into the nature of human consciousness we must next make a distinction between sense and intellect. The sensitive faculty in the more perfectly organized animals, possesses, as we judge from the arrangement of the nervous system and from results in actual life, a certain consentience, which, in the less strict sense, may be called consciousness. St. Thomas teaches that the sensitive apparatus is, after its manner, sensible of its own sensations; though what is the relation between outer organs and cerebral centres, in bringing about this effect, need not here be discussed. While the horse or the dog are incapable of the full recognition of a self as such, they have, in the inferior order, a practical appreciation of self, which ministers to their pleasures and pains, and to self-preservation. But it would be going far beyond data to argue a more perfect knowledge of self from the signs which animals exhibit of vanity or jealousy, analogous to these passions in man. As an animal, man also has his consentience, or *sensitive* consciousness. What, however, specially interests us is man's *intellectual* consciousness, which some scholastics subdivide into *direct* and *reflex*. In a broad sense all consciousness, so far as it includes some knowledge of the subject-knowing, some return of self upon self, must be *reflex*: still the difference here intended will appear in a simple example. While we explicitly perceive the truth of a geometric principle, we implicitly, in the same act, *in actu exercito*, are made aware of our own knowledge. This is styled *direct* consciousness.

X

Afterwards, by a new act, of set purpose, *in actu signato*, we may return upon our late perception, and make this, the mental fact, the object of our explicit knowledge. This is styled *reflex* consciousness, as being expressly reflex. In the one case, while we know an object we are subordinately conscious of our knowledge; in the other case, we make this consciousness the principal matter of our reflexion, and degrade the object to a subordinate place.

It will render a man all the more cautious in denying the possibility of the direct consciousness, if he considers how, in its absence, it becomes apparently impossible to have the reflex consciousness. A thought not in consciousness is, in itself, a sufficiently difficult notion to entertain; but supposing such a thing to have had place in us, how are we ever to recover it by means of what we have called reflex consciousness? How is memory to catch up an act which is bygone, and which, while it lasted, never was in immediate consciousness? Sometimes, indeed, by simple inference we may gather that a certain idea must have passed through the mind, though we have no recollection of the fact. But the more we think of it, the less will inference be deemed capable of supplying the want of all direct consciousness. With Father Palmieri,[21] who understands by *sensus intimus* what we have called *conscientia directa*, we may argue thus: "The act of the innermost sense (*i.e.*, of direct consciousness) is not in reality distinct from the act it reports, or at

[21] *Logica Critica*, thesis xi.

most the distinction, if any, is a mental distinction; in other words, when the living agent feels (is conscious of) his act, his experience of it is in reality nothing else but the self-same act objectively present to the thinking, feeling, or appetitive agent. For if another act were needed, this in its turn would have to be reported by direct consciousness, which is then supposed to be distinct from the first act. This second act, for the same reason, would have to be taken as a distinct act from a third, and thus we should require an endless series."[22] In other words, if our acts of knowledge did not at once link themselves on to a conscious self, they never could become so attached at all.

It is with the fullest advertence to the difficulty we have in "numbering off" acts of mind that the last pages have been penned. It is only very roughly that we designate a process to be one act in material operations, and when we get beyond these, and ask ourselves how many acts the mind can or does perform at once; whether there is succession between acts or contemporaneity; whether a given result requires one act or more; undoubtedly we are on ground which is to us generally very obscure. It is the teaching of St. Thomas,[23] that the mind can exist in only one state

[22] "Actus sensus intimi non distinguitur realiter ab actibus qui sentiuntur, sed tantum ratione: scilicet cum vivens actum suum sentit, experientia hæc sui actus non est realiter nisi ipse actus objective præsens sentienti vel cogitanti vel appetenti. Si enim alius actus requireretur, hic quoque rursus sentiri deberet sensu intimo, qui ab illo primo actu supponitur distinctus, eadem ratione ab hoc etiam distinctus dicendus erit, et sic ibimus in infinitum."
[23] *Summa*, Part I. q. lxxxv. art. iv.

at a time, and that an apparent multiplicity of simultaneous acts must really be reducible to a unity. There is always a great difficulty in discussing such subjects in detail, because we can form no picture to ourselves of the mode of operation proper to a spiritual substance, which has not separate parts, but which works with a marvellous unity and simplicity. To overlook these truths would lay us open to the danger of multiplying, or refusing to multiply, acts, in a way which reason could not afterwards justify. An analysis into mentally distinct parts does not prove physical parts. Hence in the little that has been said about direct and reflex consciousness, care has been taken to speak within the bounds of legitimate analysis. Of the direct consciousness, which is the most difficult to speak about safely, rather than say that one act is conscious of outer object and of inner self, we say that one act, whatever its simplicity or complexity, suffices to constitute the mind conscious of outer object and of its own knowledge of that object. Thus we make the subject of the predication rather the mind acting, than simply the act. At least this is a safer form of wording.

With a still further view to being safe in the form of wording, we may note the special difficulty which, when we are dealing with an act of volition, lies against saying, that in direct consciousness the act of volition becomes part of its own known object. For we do not attribute knowledge to the volition as such: rather we speak of will as enlightened by intellect, and of intellect as cognisant of volitions. Hence with regard to our immediate consciousness of the acts of our will, we are led to devise this mode of expression, that, without further determination, the mere presence of the volition in the soul suffices to enable the intellect to be simultaneously conscious of its presence, while some intelligence of an object is the pre-requisite of any volition at all. How, moreover, the distinction between intellect and will can in any sense be called real, is discussed in psychology: at least the soul as knowing may be distinguished from the soul as willing. Another cautionary remark is that while we have not been talking in comfortable oblivion of the difficulty which besets the numbering and the distinction of mental acts and faculties, so neither have we been oblivious of the very strong objection which some philosophers have to the idea of thought or self becoming an object to itself. Mr. Sully is but following Comte, Dr. Maudsley, Mr. Spencer,[24] and several others, when he affirms that all introspection must be retrospection: that man can reflect, not on the mental state which *is*, but only on that which *was*. In reply we must be allowed to plead, that this is reducible in the end to an *a priori* dogma, or at best to a false analogy taken from material action, and is refuted by facts. Do what we will, we cannot be true to fact and deny a real reduplication, as it were, of thought upon thought and of self upon self. There is in us a power of genuine reflexion. The mind has a

[24] *First Principles*, Part I. c. iii.; *Psychology*, Part II. c. i., The Substance of Mind.

re-entering, self-penetrating, self-permeating activity which makes it more intimately at home with itself than anything which a materialistic philosophy can allow; and, say we, all the worse for materialism, not for facts. So far, however, as the denial that thought can become object to itself, rests on an author's definition of "object" or "self," we can only beg him to improve his definitions, and allow for that marvellous gift of self-consciousness, which we all undoubtedly possess, but which recent definitions seem expressly devised to exclude.

(c) Now that we have first called attention to the fact of sensitive consciousness and next considered intellectual consciousness in its two branches, direct and reflex, we must give a moment's attention to the relation between sense and intellect in respect to consciousness. A man's feeling of hunger, for example, does not stop short at its animal level, but the subject becomes intellectually conscious that his stomach is craving for food. Thus we are reminded that one object of our intellectual perception is our own bodily state: and because, by our definition, the body and its affections are part of the composite self, such perceptions must be ranked under the category of self-consciousness. Hereupon a question suggests itself. We have been unable simply to accept the fact of an unconscious sensation or of an unconscious thought; but may not there be some sensations, present indeed to the sensitive consciousness, but never manifested to the intellectual consciousness: so that we can never intelligently affirm their presence? To answer, Yes, might to some sound an unproveable assertion, but at any rate it would, as a proposition, not contain that intrinsic conflict of terms which we seem to see in the affirmation, that some thoughts are unconscious. In any case, there are many facts of sensation which become objects of intellectual consciousness, and this relation between the two departments of consciousness is the point to which we have been directing attention.

(*d*) We are now in a condition to propose our own classification of the objects of consciousness, and to defend the validity of consciousness in their regard. To begin with the affections of the composite self, we have bodily affections, cognitive and appetitive, as sensibly perceived; we have the same again as intellectually perceived; and thirdly we have the spiritual affections, cognitive and appetitive, of course intellectually or spiritually perceived. All these objects are connected with our own person. Next, so far as whatever outer objects we know, or have any volition about, are known at least in some reference to our conscious self, this element of self, again making its appearance, justifies a use of the word consciousness, whether we distinguish it from what is more rigorously self-consciousness or not. Thus the term consciousness, as Hamilton[25] in one place declares, is ultimately extended to our whole sensitive and intellectual life, so far as we are rendered aware of our condition, whatever the object of cognition or appetency.

[25] *Hamilton's Reid*, Note B, p. 810.

Against the above classification a difficulty of minor importance might be raised. Mr. Bain dislikes calling all our emotive states by the name of volitions, because he surmises that there are some neutral feelings, in regard to which we have no appetencies either for or against them. But really to make provision for such vague and disputable states, it is not worth while disturbing the old division into cognitive and appetitive powers, whether of sense or of intellect. Accordingly no scruple has been made about going on the lines of the old tradition.

A short exposition of facts will now establish the principal thesis of this chapter, that consciousness cannot but be valid in what it testifies about self, as also in what it really testifies about non-self, so far as it may be applied indirectly to this latter region. It is a position the very sceptics have been unable to impugn, that facts of consciousness, as such, cannot be gainsaid. Whatsoever a man is conscious of, of that he is conscious; and this principle must be extended to the feeling of certainty about any objective truth, no matter what, which is presented to the mind with objective evidence. As Mr. Conder argues: "Since the presentments of consciousness are not judgments but primary facts, they cannot be unreal: only our interpretation of consciousness may be erroneous." "On this," adds a critic in *Mind*, "we are all agreed." The matter may be brought under a larger doctrine propounded in Part I., chapter ii., that no mere apprehension can be other than true, however erroneous may be the judgment of which it is made the occasion. As a case in point, what a man with an amputated leg feels, he really must feel; but he judges amiss when he declares the feeling to be in a member which he no longer possesses. The like may be said of a fever patient who complains of being cold; of the Arctic explorer who, touching a piece of long-exposed iron, pronounces it hot; of the man who says that he has the experience as of two selves contending within him—a pathological state, which, it is surmised, may be due to some want of co-ordination between the two hemispheres of the brain. So far, however, as any insanity creeps in, the subject is no longer fit to serve as a specimen of normal humanity.

Still it may be urged, if the interpretation of conscious facts may be wrong, how are we advanced beyond idealism by the assurance, that at least we may rest secure as to the facts themselves? We do not allow that interpretation is so liable, at all times, to error, that never can it be safely trusted. It is guaranteed by the conditions already stated in the chapters on the Criterion of Truth, on Error, on the Veracity of the Senses, on the Validity of Ideas. All that the present chapter adds to what has gone before, is a clearing up of notions upon what is meant by consciousness, and an emphasizing of that truth, so neatly stated by Cousin: "It is an inherent attribute of reason to believe in itself."[26] The root of agnosticism is an unreasonable distrust of reason in itself, as the root of sound philosophy is a legitimate self-confidence on the part of the

[26] "C'est un attribut inhérent à la raison de croire à elle même."

mind in reliance, upon its conscious powers.[27] As will appear in Psychology, it is rather to the right reading of consciousness that we must appeal, than to a theory about the dynamics of motives, when such grave questions have to be settled as that of the freedom of the will; and the same holds true of many other philosophical questions, notably about consciousness.

Not at all, therefore, can a special chapter on Consciousness be deemed superfluous.

### ADDENDA.

(1) Kantians have got such a decided position in this country, that their leader's theory on consciousness ought not to be quite passed over in silence: though we must beg leave to reject it on the ground that it is against the immediate light of evidence, resting as it does on a denial of the facts that some of our clearest conceptions of things are more than forms of the mind, and stand for objects which the mind can contemplate as such. Kant then distinguishes the *empirical consciousness* which takes note of the changeable conditions of the subject, from the *pure consciousness* which is *a priori* and unchangeable:

---

[27] Kant's doctrine on the necessary illusions of the reason, of which we have spoken before, certainly goes along with affirmations on his part that the faculties themselves are infallible, and that the illusions of reason are as corrigible as are the illusions of sense, such for example as that whereby the moon appears larger on the horizon. To this extent Kant is to be acquitted of the charge of making reason essentially erroneous. It is to the judgment that Kant attributes error; and though we have seen Rosmini (Part I. c. ii.) quoting Kant as an author who makes judgment the one fundamental act of Understanding, we must remember that *Understanding* is not here co-extensive with the whole mind, but is distinguished from sensitive intuition on the one side, and from reason on the other. (Max Müller's translation of the *Critique*, pp. 60—70.)

but he utterly denies that we can be conscious of a substantial *Ego* or personality. "It is clear," says Kuno Fischer, "that the thinking subject can never be an object of possible knowledge, because it is merely the formal condition of possible knowledge; and it cannot be an object of intuition, because it forms in itself no phenomenon, but only the highest formal condition of phenomena. All the conditions are wanting for us to judge that the subject of thinking is a thinking substance, or that the soul is a substance." Again: "The *Ego* is no object, but only appears to be one: it is the formal logical condition of all objects. On this illusion rests the whole of rational psychology: *I think* does not mean *a substance thinks*. That I am conscious in all my various states of my unity does not mean that a substance is conscious of its unity—that there is a personal substance. From the mere *Ego*, torture it as you will, you can never prove an existential judgment. From the mere unity of self-consciousness there follows no cognition of any object. That in all my states I am conscious of my subjective unity is a mere analytical judgment, which brings us no further than *I think*." [1]

(2) The curious may find some interest in seeing how theorizers of the calibre of Hartmann work out the notion of unconscious thought. It is impossible to say exactly what that author means; but he has some such fancy as that the Great Unconscious evolved the universe for a long time intelligently, but without consciousness. When at last a sudden shock produced consciousness, this was found to be a source mainly of pain. Hence the desirability of bringing about the abolition of consciousness.

[1] Professor Mahaffy's translation of K. Fischer on Kant, pp. 179 —185. Cf. Max Müller's translation of the *Critique*, Vol. II. p. 347.

(3) The subject of latent thought is one into which we cannot probe very deep. What is styled our *habitual*, as distinguished from our *actual* thought, is certainly something permanently existing, even while we are not using it. The historian with the materials of half a Record Office stored up in his memory, whilst he is wholly engrossed with the one thought of his own money affairs, indisputably keeps his knowledge in a latent condition; though how to describe this condition is to us a great puzzle. If "unconscious thought" is a phrase used to express this undoubted fact, then it has a true significance. But more often it signifies operations going on, with rational results, among the hidden material, or even additions made to it by fresh observations, and then the question becomes more intricate, and many are inclined to suppose some degree of consciousness to enter in, scarce noticeable at the time, and straightway forgotten—evanescent as a dream, the memory of which is occasionally preserved by the merest accident, but generally quite lost.

(4) The assertion of our consciousness about our own ideas, if clumsily made, is just what gives the appearance of the error we have so strongly repudiated, namely, that knowledge is of ideas, and that to get from ideas to things requires a bridge which no philosopher can build.

(5) As bearing out the statements in the text about the complexity of human action and the difficulty of numbering acts, a report of an address by Sir James Paget, delivered at the Mansion House, March 4, 1888, is worth preserving. "He remembered once hearing Mdlle. Janotha play a presto by Mendelssohn, and he counted the notes, and the time occupied. She played 5,595 notes in four minutes, three seconds. It seemed startling, but let them look at it in the

fair amount of its wonder. Every one of those notes involved certain movements of a finger—at least two: and many of them involved an additional movement laterally as well as those up and down. They also involved movements of the wrists, elbows, and arms, altogether probably not less than one movement for each note. Therefore there were three distinct movements for each note. As there were twenty-four notes each second, the total was seventy-two movements per second. Moreover, each of these notes was determined by the will to a chosen place, with a certain force, at a certain time, and with a certain duration. Therefore there were four distinct qualities in each of the seventy-two movements in each second. Such were the transmissions outwards. And all these were conditional on consciousness of the position of each hand and each finger before it was moved, and, while moving it, of the sound of each note, and of the force of each touch. All the time the memory was remembering each note in its due time and place, and was exercised in the comparison of it with other notes that came before. So that it would be fair to say there were no fewer than two hundred transmissions of nerve force outwards and inwards every second; and during the whole of the time, the judgment was being exercised as to whether the music was being played worse or better than before, and the mind was conscious of some of the emotions which some of the music was intended to impress." An appeal to the word *automatism* will not dispel the marvel of th performance.

# CHAPTER VI.

## MEMORY.

*Synopsis.*
1. Definition of memory.
2. The veracity of memory, and how far it can be made matter of proof from experience.
3. Limited power of memory.
4. Freaks of memory no disproof of the normal faculty.
5. Memory contrasted with anticipation.
6. Incidental use of the fact of memory to refute pure empiricism and rigorous idealism.

*Addenda.*

1. It would be a fatal thing for us if we had not what is sometimes called mental adhesiveness, that is, if nothing which we learnt "stuck." But we all recognize a power of retentiveness. Though the amount of knowledge which, at any one time, is actualized in the mind may be small, yet, below the surface of consciousness, and, under many limits, ready at call, is a comparatively vast mass of gathered information, and of skill in its use. Some writers, after Aristotle, distinguish a storing power (μνήμη) from a subsequent recalling power (ἀνάμνησις); but it will be enough for us to include both under the one name, Memory, habitual and actual.

This memory is not so clearly defined a term,

even in its wider usage, as we might imagine. A person is rather loth to say that he remembers a road which he is taking almost every day of his life, or the meaning of a word told him only a minute ago. The definition of memory ought to include two elements, the recalling of the past, and its recognition as past. To begin with the first element: if a new thought is sustained in the consciousness for five minutes, we may agree—and it is partly a matter of agreement—not to call this memory. But if a thought is allowed once to sink below actual consciousness, and then is resuscitated, no matter how speedily, we may call this Memory, so far as it fulfils the requirement of a recalling of the past. In practice, however, it is often impossible to say whether we have momentarily let go an idea or not; and furthermore, when the interval is very small, as there is no sufficient test of retentive power, men seldom care to distinguish such a revival from the first impression or conception. The second element of memory has its absence illustrated by the man who honestly repeats his friend's epigram or joke as his own, wondering the while at his own readiness of wit; or again, by the old person whose memories are mistaken by him for present circumstances. Though, however, we distinguish remembered from fresh knowledge, we should bear in mind that the adult never discovers anything altogether new: his fresh acquisitions always combine together with a great many old stores. We should at once feel the puzzle of locating an entirely new fact; and in general we may safely affirm of every adult man, that every act of knowledge which he elicits must be largely made up of memories.

2. The veracity of memory, as a general faculty, is made intuitively evident during the course of its use. Even Mill was driven to allow, that we must put an intuitive trust in our power of reminiscence, though he forbore to make the handsome acknowledgment that in so doing we are not blindly instinctive. Again, Dr. Ward,[1] in his passage of arms with Mr. Huxley, was undoubtedly triumphant over the Professor when the latter undertook to show, that the validity of memory can be proved empirically by successive trials, overlooking the fact that for the knowledge of the success of these repeated experiences he was relying all the time on memory. Yet we all must admit, it is only in experience that memory shows its powers, and brings them home to consciousness; there is no *a priori* revelation of its trustworthiness. Allowing a certain intuitive perception of the validity of the faculty given in its first exercise, a man can then go on to test the extent of its ability; and he can confirm his confidence by sundry experiments not difficult to devise. In like manner, it is empirically that some men learn that their memory is very deficient, or has *lacunæ* in it. At times a fact which we directly remember may be further verified by calculating back from present data, and proving that the fact must have been as remembered. These admissions may safely be made about the possibility of putting memory to the proof: all the same,

[1] See the Preface to his *Philosophy of Theism*.

the ultimate guarantee for the validity of the faculty is the immediate evidence brought forth in the exercise of remembering; and this is implied in all our proofs.

3. While, however, memory is undoubtedly a valid faculty, its limited character is equally beyond a doubt. It may fail in either of its branches, either that of recalling or that of recognizing. For practical purposes much that we have once learnt is lost as explicit knowledge; and many facts are so vaguely recollected, that we do not know whether to call them reminiscences or imaginations. The important point in connexion with the weakness of memory, is not to be deceived into taking it as an argument for the radical incapacity of the instrument, but to take it rather as a warning to improve an imperfect faculty by cultivation, and not to spoil it by abuse. Much may be done by orderliness, by strict truthfulness, by careful discrimination of facts from fancies, or prejudices, or desires, and by distinguishing when it is that we clearly remember, and when it is that we are perplexed. Any ordinary man would feel that his life was safe if it were simply staked on his correctness in enumerating one hundred facts of memory at choice: while he would feel great alarm if the one hundred facts were assigned by another, and belonged just to the region where memory began to grow shadowy. A third hundred of events might be assigned which would simply make him despair. Therefore, a real power with limitations—such is the description of human memory.

Y

4. What inclines some people to speak ill of memory as a faculty, is the very great and striking variety of its abnormal conditions. Diseased or declining state of mind often shows its beginnings by the manifestation of injury done to the power of remembrance. A man's consciousness may be split up into two or more almost completely isolated series, which cannot be brought into union with each other. A person lays in a stock of knowledge during a number of years, then he has a sickness, which leaves him under the necessity to begin the learning process over again; next, he may suddenly relapse into his first mental condition, and after that, alternate between the two states. Again, a patient may forget all the words beginning with certain letters, or the whole of one language; or he may recognize the spoken, but no longer the written word, though he sees it. Thus memory may fail in departments. Others again more and more lose the discrimination between things remembered and things only fancied. All which proves, indeed, how frail and liable to frustration is memory, depending as it does on the preservation of very complex organic conditions: but as long as a man keeps his mental sanity, he can take account of his pathological state, and make allowances for recognized flaws in memory. Some have bravely done this with success under painful circumstances. Cases of aphasia furnish occasionally good illustrations. But unfortunately with disease of the memory there often goes a general disease of the reason; and then the victim is no longer fit to serve

as a standard man, from whom to take the measure of the human memory. He cannot become even to himself a disproof of that faculty; for such disproof must always fall into the old vicious circle of disbelieving a faculty in reliance upon the faculty—a process not so feasible as setting a thief to catch a thief.

5. The subject of memory receives further light from a comparison with the faculty of anticipation, to which it is sometimes too closely likened. Apart from extraordinary processes of foresight, which do not concern us, there is no faculty of immediate anticipation corresponding to what may be called immediate memory. Such a faculty would be quite unaccountable, whereas of memory an account can be given. Impressions abide till positively effaced even in material things: and we are ready to expect that impressions should abide also in the faculties of knowledge. Moreover, the impressions of knowledge were received in a certain order, and of this fact also a trace may fairly be expected to remain. In reliance upon it, we sometimes mentally trace back a fact, link after link, in a chain of associations. What seems immediate memory may be something like the instantaneous retracing of these steps, or at least of some of them. But for anticipation we have no such mental residua to fall back upon; for the experiences are yet to come. Hence at the very most we can reason out a future event from present data, just as we might reason out a past event, which we had not perceived as it past, and therefore could not recall by memory.

6. An incidental use of the fact of memory is, that its inevitable admission is fatal to pure empiricism, and to the pretence of rigorous idealism never to transcend the fact of present consciousness. For first, as we have seen, memory must, at starting, demand an intuitive trust in itself, and can never be guaranteed simply by an inference from accumulated experiences. Secondly, to allow that we know any fact as really an event of past time, is to give up the idealist dogma that no idea can travel beyond its own bounds to an object not itself. Thus our previous conclusions receive incidental confirmation, and the theory of adversaries has to submit to one more exhibition of inconsistency. Faculty is proved to be not simply the product of function; only a previously existing faculty can develope itself by functioning. Memory is not the creation of experience, but it manifests itself, grows, and is perfected by experience. And memory, whether primitive or highly developed, always transcends the present, and refutes the first principle of thorough idealism.

### ADDENDA.

(1) Reid[1] has a passage calculated to give rise to some controversy: "I think it appears that memory is an original faculty, given us by the Author of our being, of which we can give no account, but that so we are made. The knowledge which I have of things past, by my memory, seems to me as unaccountable as an immediate knowledge would be of things to come; and I can give no reason why I should have one and

[1] *Intellectual Powers*, Essay iii. c. ii.

not the other, but that such is the will of my Maker." That memory is more intelligible than foreknowledge has been already argued in the principal text: and against calling knowledge of either past or future, *immediate*, Hamilton, after the requirements of his theory, enters a protest in a note. But, leaving these points, we may turn to another, and ask in what sense is memory a *peculiar* faculty? Here it looks as though a caution were needed against the double extreme of making memory too much, and of making it too little peculiar. The intellectual memory is one faculty with the intellect, and yet it is a special exercise of that faculty,[2] the peculiarity of which should not be overlooked. Against such oversight Sir H. Holland makes the remark:[3] "We do not gain greatly from these metaphysical definitions, which resolve memory altogether into other phenomena of mind. Among modern writers on the subject, Dr. Brown has gone furthest, perhaps, to merge this faculty in other functions and names." The pith of Brown's doctrines seems to be conveyed in the following sentence:[4] "To be capable of remembering, in short, we must have a capacity of the feelings which we term *relations*, and a capacity of the feelings which we term *conceptions*, that may be the subjects of the relations: but with these two powers no other is requisite—no power of memory distinct from the conception and relation which that complex term denotes." The relation he explains to be one of priority and of subsequence between concepts. Hamilton agrees with Brown so far as to maintain that memory is quite an explicable function of the intelligence; and the precise point needing explanation he makes to be the persistence and the recognition of past intellectual

[2] St. Thomas, *Summa*, Part I. q. lxxix. a. vii.
[3] *Chapters on Mental Physiology*, c. vii, p. 149.
[4] *Human Mind*, Lecture xii.

acts: "I think we can adduce an explanation founded on the general analogies of our mental nature."[5] For the retentive part he borrows the account of H. Schmid: "The mind affords in itself the very explanation we vainly seek in any collateral influences. The phenomena of retention are indeed so natural on *the ground of the self-energv* of the mind that we need not stop to suppose any special faculty for memory; the conservation of the action of the mind being involved in the very conception of its power of self-activity. It is a universal law of nature, that any effect endures as long as it is not modified or opposed by any other effect. But mental activity is more than this; it is an energy of the self-acting power of a subject one and indivisible; consequently a part of the *ego* must be detached or annihilated, if a cognition, once existent, be again extinguished. At most it can be reduced to the latent condition." After so accounting for Retention, Hamilton accounts for Reproduction or Resuscitation by the laws of Association, which he thinks make abundantly clear what was so obscure to the scholastics, that Oviedo called it "the greatest mystery of the whole of philosophy."[6] In materialistic phraseology, Dr. Maudsley[7] describes memory as extending analogously throughout organic life: "There is memory in every nerve-cell, and indeed in every organic element of the body. The permanent effects of a particular virus on the constitution, as that of small-pox, prove that the organic element remembers, for the rest of life, certain modifications which it has suffered; the manner in which the scar on a child's finger grows as the body grows, evinces that the organic element of the part does not forget the impression that has been made upon it. The residua by which our

---

[5] *Metaphysics*, Lectures xxx., xxxi.
[6] "Maximum totius philosophiæ sacramentum."
[7] *The Physiology and Pathology of Mind*, c. ix. p. 209.

faculties are built up are the organic conditions of memory." What Dr. Maudsley does not labour to explain is, the passage from organic conditions to intellectual memory.

(2) A continuation of the last quotation will lead us on to Mr. Spencer's theory of memory: "When an organic registration has been completely effected, and the function of it has become automatic, we do not usually speak of the process as one of memory, because it is *entirely unconscious*." The last phrase is disputable: and in all cases we must protest against memory being set down as a mere transitional stage on the way to bodily automatism. It is constantly the tendency of Mr. Spencer's doctrine to regard the automatic adaptation of organism to material environment as the highest goal; and all stages in consciousness as so many accidents by the way, to be got rid of by higher development. Such is the tendency of his doctrine on Memory:[8] "So long as the psychical changes are completely automatic, memory, as we understand it, cannot exist. There cannot exist those irregular psychical changes seen in the association of ideas. But when, as a consequence of advancing complexity and decreasing frequency in the groups of external relations responded to, there arise groups of internal relations which are imperfectly organized and fall short of automatic regularity, then what we call memory becomes nascent. Memory comes into existence when the involved connexions among psychical states render their successions imperfectly automatic. As fast as these connexions, which we form in memory, grow by constant repetition to be automatic, they cease to be part of memory. We do not speak of ourselves as recollecting relations which have become organically registered. We recollect those relations only of which

[8] *Psychology*, Part iv. c. vi. § 200, p. 445.

the registration is incomplete. No one remembers that the object at which he looks has an opposite side; or that a certain modification of the visual impression implies a certain distance; or that the thing he sees moving about is a live animal. To ask a man whether he remembers that the sun shines, that fire burns, that iron is hard, would be a misuse of language." Nevertheless these several items would come under the head of memory, as we have defined that term; nor should we admit, that "the practised pianist can play while his memory is [wholly] occupied with quite other ideas than the memory of the signs before him"—if indeed he is playing from the signs as his guides. In conclusion Mr. Spencer thus describes the transitional character of memory: "Memory pertains to that class of psychical states which are *in the process of being organized*. It continues as long as the organizing of them continues, and disappears when the organization is complete."

(3) M. Ribot, whose doctrine, in his volume on *The Diseases of Memory*, is that "memory is *per se* a biological fact, by accident, a psychological fact," discusses the position of the latent results stored up in memory, and waiting to be called into actual use. He thinks our best course is to describe these *residua* as "functional dispositions," not as in any way conscious acquisitions; for "a state of consciousness which is not conscious, a representation which is not represented, is a pure *flatus vocis*." Hence he asserts "a minimum of conscious memory" in those who "are able to rise, dress, take meals regularly, occupy themselves in manual labour, play at cards and other games, frequently with remarkable skill, while preserving neither judgment, will, nor affections." He might in these cases allow some degree of judgment and will, over and above pure unconscious automatism.

# CHAPTER VII.

### BELIEF ON HUMAN TESTIMONY.

*Synopsis.*
1. Belief on testimony is a special subject, calling for special treatment.
2. Naturalness of such belief, both from the knowledge and the veracity of the speaker and from the expectations of the hearer.
3. Testimony is one undoubted source of certitude, and a very abundant one.
4. Single and cumulative witness.
5. Points on which we must be guarded. (*a*) We must distinguish the completely from the partially feasible in history, and remember that much history does not rise above probability. (*b*) A wrong point of view may disturb a whole body of facts. (*c*) Fallacy of excessive reliance on internal evidence, especially where the reader tries to impose his own circumstances on a writer in quite other circumstances. (*d*) Fallacy of the argument from silence.
6. Providence in history.

1. WHILE it is clear that the veracity of the senses, as has been shown before, forms part of the problem of our belief in the testimony of other men, it is equally clear that it is not the whole problem. There is something special about our trust in the word of another, which calls for a separate treatment. Hence it is unsatisfactory to find the elder Mill arguing thus:[1] "Belief in events or real

[1] *Analysis*, Vol. I. c. xi. p. 382.

existences has two foundations; first, our experience, and second, the testimony of others. When we begin, however, to look at the second of these foundations more closely, it soon appears that it is not in reality distinct from the first. For what is testimony? It is in itself an event. When, therefore, we believe anything in consequence of testimony, we only believe one event in consequence of another. But this is the general account of our belief in events." Yes; and still things which agree in being events may differ in being events of a specifically different order; and such is the case in the present instance. Manifestly belief in testimony has its peculiar nature, not a little important to a Christian, whose religion is historic and rests on historic foundations.

2. Belief in testimony is natural, and natural on its two sides. First, man being intelligent, is apt to discover truth, and, apart from extrinsic reasons, is inclined to declare the truth as he knows it. Even the downright liar, according to James Mill's estimate, for one lie that he utters tells a thousand truths. Secondly, on the side of the recipient, he has been accustomed from childhood to depend on the information of his elders, and from his knowledge of himself judges what he is to expect from others. Mr. Bain, therefore, seems to be throwing a needless mystery over the case, when he talks of "a primitive credulity in the mind," from which he derives "belief in testimony," and which he describes as "a primitive disposition to receive all testimony," till sad experience of deception

gradually modifies the too ready instinct. Of course, children are simple and credulous, but the appeal to "a primitive instinct" is hardly necessary to account for the fact.[2]

3. That testimony, oral and written, is a source, and an abundant source of certitude, cannot, in concrete cases, be plausibly gainsaid. The whole plausibility lies in keeping to the abstract, and is well illustrated by Mr. Balfour's ingenious arguments against the theoretic trustworthiness of any old manuscripts.[3] The author's subtleties are telling enough, when the concrete circumstances are not at hand whereby to put a rude stop to their light and airy play; but take them out of the air, weight them with the load of terrestrial facts, and straightway their frolics are over. It is simply demonstrable by way of testimony, that Alexander of Macedon and Julius Cæsar were successful leaders of armies, and produced notable effects in the world's history; also that Demosthenes and Cicero were powerful in speech; and that there was a writer of comedies called Aristophanes, specimens of whose work we yet have. In the history of our own country there have certainly been a Roman, and an Anglo-Saxon, and a Norman conquest. For in regard to these events we may be sure of the knowledge and of the veracity of the witnesses, just the two requisites for trustworthy testimony. And if we test the sceptical generalities which are urged

---

[2] *Deductive Logic*, Introduction, n. 17, p. 12; *Inductive Logic*, Bk. VII. c. iii.

[3] *Defence of Philosophic Doubt*, c. iv. pp. 53, seq.

against the possibility of any historic certainty, by instances like the above, the adversary will produce little impression, when he argues, in the abstract, that the occurrence of a fact is only one out of several equally possible causes for its assertion; that a tradition grows weaker with every transmission through a new channel; that the original force of an authority becomes dissipated among its countless recorders, and that each witness being fallible, so are any number of witnesses. Without further argument, therefore, we may take the proposition as established, that certitude in reliance on testimony may often be had. When had, it is what we have called "moral certitude," in the sense that it reposes on a knowledge of the actions of moral agents, or men, in speaking the truth. It is real certitude, though not metaphysical; and so Mr. Mahaffy, in the Introduction to his *Prolegomena to Ancient History*, is granting all we contend for, when he declares that historic belief may be beyond all doubt, but can never reach mathematical demonstration. If it is beyond all doubt, it is quite certain, and that is all for which we stipulate.

4. Unquestionably a number of independent witnesses are often required to establish an event; and the special force of the argument then lies, not only in the fact that it is unlikely so many together should be guilty of a lie, but also in the impossibility that they should have succeeded in lying consistently. When a number of witnesses are agreed to perjure themselves in a court of law, about the only safe

way to secure uniformity in the narration of a fictitious event of some complexity, is to enact the scene before the eyes and ears of all. Very impossible is it that without any previous arrangements writers should independently tell one intricate story; and sometimes it can be proved, not only that there was no prior conspiracy, but also that such conspiracy would have been ineffectual, because of other modes of information outside the circle of the presumable conspirators. Busy with these considerations about the value of a multiplicity of vouchers, sometimes people are led into the assertion that never can a single witness be a sufficient authority for a certain assent. Without entering into detail, we may protest that this declaration, in its universality, is a calumny against human nature.

5. So much in general about belief on testimony; in particular some cautions are needed for the guidance of our judgment—cautions which may be illustrated, but not exhausted, in the following observations:

(a) We must distinguish the quite feasible in history, from the partially feasible. If the historian binds himself to put down nothing but what he can fairly conclude to be beyond all controversy, he will be very meagre and dry. He will be cut off from most of what is called the philosophy of history, and reduced almost to the position of a chronicler. Such safe but jejune writing is liked neither by authors nor by readers; and hence it becomes necessary for both sides to recognize that large portions of history, as now composed, do not rise above probability. Consequently counter probabilities must be treated with the respect due to them, not as though one party were entitled to the monopoly of conjectural interpretation. Even the most probable account need not be the truest. It is to be feared that not sufficient allowance is made for the essentially problematic character of much historical writing. Hence, just as when we were considering physical certitude, we distinguished the safer from the more venturesome attempts, so in considering moral certitude we must make a like distinction. And in the category of the venturesome we should place most books which treat of comparative mythology, comparative religion, the origin of social institutions, and such matters, in which documents are scarce or obscure, or written in a language ill understood, while inferences are often marked more by ingenuity than conclusiveness. Sobriety of judgment in these subjects characterizes rather the dispassionate reader of rival systems than enthusiastic partisans.

(*b*) Another thing to note is how wonderfully a man with a point of view, especially if he is selective in his incidents, or even inventive, can make facts conform to that point of view, without at all proving that he is right. A glaring instance is Draper's *Conflict of Science and Religion*, a book which it is well to quote as an example, because it has had a wide circulation, and has done much harm to the cause of truth. Draper may have been quite honest, as honest as he declares himself to have been; but at any rate he has a wonderful power of making his point of view tell upon facts, instead of *vice versa*. Let me illustrate this power by a single but fairly chosen instance, which, in this country, will be more easily appreciated than other ecclesiastical events which have been not less misrepresented. Fancy a man, who in the light of modern research, could categorically assert without the shadow of a qualification that, "a conviction that public celibacy is private wickedness *mainly* determined the laity, as well as the government in England, to suppress the monasteries." This example will do to illustrate the force of "point of view," and its influence on Mr. Draper's credibility.

(c) A third danger is excessive reliance on what are called "internal evidences," a danger all the greater when a critic insists on carrying his own times and circumstances into distant and differently situated ages. The full bearings of this remark can be appreciated only by the actual examination of cases in point; but at least its general drift may be made intelligible. Where we are abstract, metaphysical, or literal, other people have been concrete, pictorial, or metaphorical. The unity, the sequence, and the completeness which we, as a matter of course, try to give to a historical narrative, they never dreamt of giving; but they were fragmentary, logically and chronologically "non-sequacious," and without pretence to adequacy. To mention only one instance out of several, the reticences of the Old Testament are many and manifest, especially on points of mere secular detail. How garrulous old Herodotus would have been, if he had known as much about Egypt, Nineveh, Babylon, and Persia as the sacred writers must have known; yet their remarks upon mere manners and characteristics are but incidental. There is no sketching for the sake of sketching. To suppose, then, that sacred history is something other than what it is, is to misinterpret it by judging it on a false standard. But upon so burning a question we had better stop short with what is obviously only an example by the way, rather than run the risk of damaging an important cause, by appearing to state its whole defence where no such statement is attempted.

In profane history, however, we may pursue the line of illustration already entered upon. A great error is committed by supposing old authors to have written with that completeness which is expected in our days of abundant books, of world-wide intercommunion, of accumulated results gathered from exploration in all fields, of easy means of reference to what are pre-eminently works for reference, and of recognized canons for literary production. Josephus[4] was speaking to our point when he made the apologetic remark, that it was no new thing for one people not to be acquainted with the history of another, "a fact true also of Europe, in which about a city so old and warlike as Rome, mention is not made either by Herodotus, or Thucydides, or any of their contemporaries; only late in the course of events Greece became acquainted with Rome." He adds that Greek writers knew little of

[4] *Contra Apionem*, Lib. I. n. 12.

Gaul and Spain. "How, then," he continues, "is it proper matter of wonder that our people also were unknown to many, and that a nation so separate, so remote from the sea, living after its own peculiar customs, should have given no occasion for writers to make mention of its doings?" At least there is a substantial force in this argument; and though it was the fate of the Jews to come into very rude contact with the great empires of antiquity, Egyptian, Assyrian, Persian, Greek, and Roman, yet Tacitus is a glaring example how little an intelligent historian may have known about Israel. And even with regard to their own history, the incompleteness of ancient writers is further instanced by that want of emphasis or proportion of which Cardinal Newman speaks: "Those who are acquainted with the Greek historians know well that they, and particularly the greatest and severest of them, relate events so simply, calmly, unostentatiously, that an ordinary reader does not recognize what events are great, and what events are little; and on turning to some modern history in which they are commented on, will find to his surprise that a battle or treaty, which was despatched in half a line by the Greek author, is perhaps a turning point in the whole history, and was certainly known by him to be so."

The result of this otherness of conditions in old times was, that occasionally we find just saved from oblivion an event which we should preserve in a thousand ways. In these matters instances are everything, and the following instance, as recorded

z

by Sir C. Lyell, in his *Principles of Geology*, is much to our purpose. "The younger Pliny, although giving a circumstantial detail of so many physical facts, and describing the eruption, the earthquake, and the shower of ashes which fell at Stabiæ, makes no allusion to the sudden overwhelming of two large and populous cities, Herculaneum and Pompeii. In explanation of this omission, it has been suggested that his chief object was simply to give Tacitus a full account of his uncle's death. It is worthy of remark, however, that had the buried cities never been discovered, the accounts transmitted to us of their tragical end might well have been discredited by the majority, so vague and general are the narratives, or so long subsequent to the event." What Pliny had strangely omitted nearly failed of being supplied by others, in which case his omission might easily have been taken as proof of the negative. Now, let us compare this case with an equally strange omission in more recent times, and about a more recent calamity—an omission, however, amply made up for by other sources, because the event occurred in modern times. Spinoza's friend, Oldenburg, was in London during the great plague; but, says Dr. Martineau, in his *Study of Spinoza*, "when we remember what was passing in the streets of London and on the Northern Sea during the September and autumn of 1665, it is strange to see how slight a vestige it has left on the correspondence of its witnesses or participators. In the plague-stricken city where Oldenburg wrote, ten thousand victims perished in a week; but apparently the visitation would have elicited no remark, had it not, by the interruption of business, delayed the arrival of a book, and suspended the regular meetings of the Royal Society!" Had such occasion not caused the mention, and had Oldenburg remained quite silent about the calamity, we have it, nevertheless, preserved for us in numberless other records. But in ancient times the perpetuation of such a fact might depend on a single writer, whose works were to be extant in distant time, and he might either fail to say anything, or say it so off-handedly, that the event would be either not known, or wholly under-estimated. We are warned, therefore, not to rely over much on the *argumentum ex silentio*, which some critics urge to an extravagant degree, in the case of writers who never dreamt of being exhaustive.

While we are on the subject of the differences between ancient and modern historians, the confession may freely be made, that the way in which, innocently or fraudulently, forgeries used to be committed, is very surprising to us in these modern days, and the fact much perplexes that historic truth which we wish to defend as attainable. Still an age which could so accept forgeries was also an age clumsy in the formation of them. Döllinger instances a stupid attempt to pass off some volumes at Roma as of Numa's authorship; they were supposed to have been discovered in an old stone coffin, and were written in Greek and Latin. But as paper and Greek prose were evidently articles not so readily to be had in Numa's days, the imposture was betrayed. Similarly modern criticism has been able to detect certain forgeries, though sometimes it has been too keen after a case for exposure. Neither is it first of all within modern times that any critical power has shown itself among scholars. The ancients were not all of them and altogether fools on the point, as many recorded criticisms of theirs remain to prove. In spite of many regrettable forgeries, therefore, we have a distinguishable history.

6. It is fashionable, in what claim to be enlightened circles, to ridicule Bossuet's historical compendium; but whether he has succeeded or not in tracing the providential course throughout the ages, we must bear in mind that there is a Providence in history, and even in making ascertainable to us certain vital portions of history. It would have been against the Providence of God to have allowed the two connected dispensations, Jewish and Christian, such a verisimilitude of historic support, had they been really mythical creations; or to have left them so dimly recorded that we could not substantially trace out the record. For, as one of the Fathers remarks, we might protest, "Lord, if we are deceived, Thou hast deceived us;" or on the other hand we might say, "Lord, Thou has left us without sufficient light."

ADDENDA.

(1) As an instance of the endeavour to be over-clever in historic science, we may take the case of Buckle, who so gloried in his imaginary triumph as a philosophic historian. Borrowing some ideas from others, for his conception was not new, he proved to his own satisfaction, that militaryism must die out with the advance of popular power; that wars were made by a small class, who looked to their own emolument or honour, but would not be made by the masses, whose interests were for peace. In Europe he fancied that, the popular will being dominant, we had ended the age of wars. The outbreak of the Crimean war displeased him, but did not upset his conviction. He pointed to the fact, that the quarrel originated between Russia and Turkey, two of the least advanced nations which had a footing in Europe. Littré, labouring under a like pleasant delusion, was more effectually roused from his dream; for he lived to see, what Buckle never saw, a succession of European wars, including the Franco-German, in which last grim struggle he had the poignant sense of being on the beaten side. In 1850 he had written: "Peace has been foreseen by sociology these last twenty-five years. Now-a-days sociology foresees peace for all the time to come of our present transitional state, at the close of which a republican confederation will have united the West, and have put a stop to armed contests."[1]

In 1878 his comment on the above was: "Would that I could blot out those unhappy pages! Scarcely had

---

[1] "La paix est prévue depuis vingt-cinq ans par la sociologie. Aujourd'hui la sociologie prévoit la paix pour tout l'avenir de notre transition, au but de laquelle une confédération républicaine aura uni l'Occident et mis un terme aux conflits armés."

I prophesied, in my childish enthusiasm, that there would be no more military defeats in Europe, but political defeats would take their place, than there happened the military defeat of Russia in the Crimea, of Austria in Italy, of France at Sedan and at Metz, and, quite recently, that of Turkey in the Balkans."[2] Thus poor Littré and Buckle were sadly out in their calculations; yet, reading their arguments, we find them quite up to the average plausibility, such as is to be found in recent theories of history and criticism. The course seems triumphant till it be rudely interfered with. M. Pasteur further tells us of the disappointed Littré:[3] "The work published by him in 1879 teems with the blunders into which Positivism betrayed him."[2]

(2) In his work on the *Transmission of Ancient Books*, Mr. Taylor thus speaks of the nature of old historic records: "Many instances may be adduced of the most extraordinary silence of historians, relative to facts with which they must have been acquainted, and which seemed to lie directly in the course of their narrative. Important facts are mentioned by no ancient writer, though they are unquestionably established by the evidence of existing inscriptions, coins, statues, or buildings."

[2] "Ces malheureux pages! je voudrais pouvoir les effacer. A peine avais-je prononcé, dans mon puéril enthousiasme, qu'en Europe il n'y aurait plus de défaits militaires, que celles-ci désormais seraient remplacées par des défaits politiques, que vinrent la défaite militaire de la Russie en Crimée, celle d'Autriche en Italie, celle de la France à Sedan et à Metz, et tout récemment celle de la Turquie dans les Balkans."

[3] "L'ouvrage qu'il a publié en 1879 est remplie des méprises que la doctrine positiviste lui a fait commettre en politique et en sociologie."

# CHAPTER VIII.

### BELIEF ON DIVINE TESTIMONY.

*Synopsis.*
1. Motive for adding to the philosophic account of certitude little doctrine borrowed from theology.
2. Difference between human and divine faith, when the latter is supernatural.
3. *A priori* probability of Revelation.
4. The supernatural revelation which has, in fact, been given.
5. Responsibility of writing a treatise like the present.

1. So far the claims of reason have been asserted, and put higher than this sceptical age is inclined to allow. It is just that after the assertion of the prerogatives of reason, the claims of a superior power should be briefly indicated; otherwise a false impression might be conveyed as to the all-sufficiency of man's natural lights.

2. Faith in general is belief on the authority of a speaker; and if the speaker is human, so too is the faith; if he is divine, so too is the faith, at least in some respect, but not necessarily in the degree required for salvation. For there are arguments convincing to the natural reason both as to the fact *that* God has spoken, and as to the matter, *what* God has spoken, at least so far as regards the substantial parts of His message. Reason, too,

affirms that what God says is to be implicitly received. Now, inasmuch as the revelation itself has been supernatural, this acceptance of God's word would be a faith founded partly on the supernatural; but it would not be simply what we call supernatural faith. For this further requires that the act be elicited by the co-operation of intellect and will, not as left to themselves, but as elevated by grace, and as using, for the sole motive which enters intrinsically into the very act of faith itself, the authority of God. It follows that what are called *præambula fidei*, are the suitable preparatives for the assent called the act of faith ; but they neither give to it its formal motive, nor lead by mere natural force to its being elicited. Hence the great error of those who are accustomed to regard supernatural faith as the mere outcome of reasoning upon the Christian evidences.

To repeat the same doctrine in other words. Supernatural faith normally presupposes at least some sufficient portion of the arguments which apologetics supply, and goes beyond into quite a higher sphere, into which the force of the apologetics could never raise it. In order to produce saving faith, grace, with the twofold office of enlightening the intellect and impelling the will, must enter into the soul and its powers. The mind so elevated elicits the act of belief. Thus the motive of faith, strictly so-called, is not found in the grounds for coming to the reasonable inference that God has revealed a certain truth, but in the word of God alone, in the divine authority, in the acknowledged omniscience and veracity of God revealing. "I believe this article on the divine word"—such is the formula expressive of the act of divine faith. The presence of grace in this act is not usually a matter of direct consciousness: rather it is known by the secure trust we have, that God will do what He has promised to do, if we honestly endeavour to fulfil the conditions.

Faith so regarded will no longer be looked upon as a simple matter of intellect. Seeing its supernatural character, its only partial and extrinsic dependence on the natural preliminaries, we shall the more readily admit that God supplies in the ignorant the defect of scientific apologetics; that He sustains the really faithful in their conflict with learned infidelity; that the preservation of faith once received is no mere matter of examining every fresh objection and triumphantly solving it. Knowing that while reason is somehow at the basis of faith, it is not the whole basis—that it is not simply the root out of which faith naturally grows—we shall have a truer estimate of how reason stands to faith as its condition; so that there is no faith without reason, and yet reason alone is inadequate to the production of faith.

3. Faith in revelation being as described, it is left for us to consider how readily disposed we should be to acquiesce in the providential order, that unaided reason should not for us be all in all. A revelation is *a priori* probable. Its probability is suggested by our ignorance, which is only too keenly felt. For no sane man would say, I am so clever, I am

above being beholden to the aid of a teacher. When a schoolboy shows no sign that he can be made aware of his own ignorance, then, whatever his "sharpness," our hopes for him are not great. Neither should we think very highly of any scientific man who had not realized the inadequacy of human science; who did not see that, even when we succeed in submitting physical phenomena to mathematical calculation, the mathematical aspect is but an aspect, and leaves other sides of the truth undiscovered. Mr. Tyndall represents himself as confounded with the vast mysteries left undiscovered in the universe, and as asking himself the pertinent question, Can it be that there is no Being who understands more about things than I do? Now, human ignorance, felt in matters of physical science, is a drawback, but does not touch on highest interests; whereas human ignorance felt, as the mass of men, when left to their natural resources, do feel it, about the very origin and end of their existence, certainly touches on highest interests. Hence it is *a priori* probable that the Creator has supplied, by some special communication, what He has left imperfect in our means of discovering truth for ourselves. Probably He has made an external revelation the complement of inner incompleteness. Not that we must exaggerate this latter defect, and speak as though reason were incapable of finding out man's destiny; but taking the bulk of mankind, we are safe in saying, that without revelation they have not a sufficiently easy, sure, and universally available means of keeping constantly in mind how they stand related to life, death, and after-death. Circumstances thus show the likelihood of a revelation.

4. As taught by the revelation which we have actually received, we know that in view of the strictly supernatural end to which *de facto* we are destined, revelation is not merely a matter of more convenient provision, but an absolute necessity. Not natural knowledge, but supernatural faith is the sole assent of intellect, which is now available for salvation; or, as St. Paul expresses it, *Sine fide impossibile est placere Deo*—" without faith it is impossible to please God."

5. Reason, not faith, has been the main point of defence in the foregoing pages; but in defending reason we have been promoting the cause of faith. Vilify reason, and you will never make good the title of faith to be honoured; but secure to reason her due position, and she will be able to add to her own dignity by defending the dignity of faith, and claiming to herself her due participation in this higher light.

Such being the final use of that part of Philosophy which it has been the purpose of this book to explain, it is manifest how no one, who has a sense of responsibility, can offer to the public a treatise on this subject without feeling how much his work is "stuff of the conscience." It is an awful crime, in the spirit of levity, to meddle with the springs of human knowledge; to spread abroad heedlessly doctrines that may be infinitely mischievous; to allow an itching for novelty, or the display of ingenuity, to make the pen write what the sober judgment cannot acquit of rashness; or to permit fear of being thought old-fashioned and mediæval, to dictate the adoption of what is new-fashioned and modern, and worldly-wise, yet all the time is an outrage, more or less conscious, upon the sacred cause of truth.

# INDEX.

ABSTRACTION, its use 169.
ACADEMICIANS, account of 135 ; their use of probabilities 139.
ACCIDENTS, definition of 252.
AFFECTIVE MOVEMENT, unable to be clearly isolated 66.
AFFIRMATION, first act of thought 162.
AGNOSTICISM, and possibility 95 ; its want of self-confidence 361.
AGNOSTICS pledged to scepticism 143.
ALLZUSAMMENHEIT in philosophy 116.
ANIMALS compared with man 27 ; their appreciation of self 353.
ANTICIPATION compared with memory 371.
APPEARANCES, truth of 225.
APPERCEPTION contrasted with perception 37.
APPREHENSION, definition of 14, 15 ; controversy on 15 ; never false 16 ; doubts raised by definition of 17 ; opinions on 28 seq. ; its identity with judgment 38.
ARRANGEMENT OF THOUGHTS, utility of 43.
ASSENT used for "opinion" 46 ; the *perspectio nexus* 51 ; motive of 53 ; the character of certitude 54 ; metaphysical 55 ; admits of degrees 59 ; living 62 ; St. Thomas' definition 62 ; intellectual and emotional 63 ; under compulsion 64 ; cause of its intensity 66 ; blind 189 ; an act of will 246 ; simple and complex 347.
ASSOCIATION, importance of term 74 ; Mill's opinions 74 ; of sensations 77 ; of attributes 325.
ASSOCIATION OF IDEAS, in place of evidence 91 ; cannot account for substance 254.
ASSOCIATION THEORY, exaggeration of 103.
BECOMING distinguished from "being" 177.

AA

BEING, necessary and contingent 92; distinguished from "becoming" 177.
BELIEF, the characteristic mark of Judgment 25, 26; Hume's definition 29; definition of 47; Hamilton's opinion 47; erroneous notion of 48; in the Catholic Church 48; spontaneous 75; due to association 75; persistent 171; opinions on 209.
BILOCATION not naturally possible 306.
BODIES, their attributes 282; Mill's theory of 289.
CARTESIANISM, popular idea of 149; really scepticism 153; result of 155; has no supporters 158.
CAUSALITY, theory of 86; *efficient* denied 94; its nature 254; the condition of experience 259 *n*.
CERTAINTY, *see* Certitude.
CERTITUDE, definition of 42; necessity of its admission 43; stages on the way to it 43; its position in logic 46; essential character of 54; metaphysical 55; physical 56; moral 57; difference in degree of 59; negative and positive 59; equation of 61; its variation in degree 63; intellectual and emotional 63; interdependence of the three kinds of 65; plain denial of 76; De Morgan on 101; real possession of 108; natural opposed to philosophic 119; analysis of motives of 122; proportionate to its known motives 123; to be preceded by doubt 151; proposed criteria of 188; supra-sensible 190; not destroyed by unsolved difficulties 229; views of 256; scientific 347; moral 380.
CHANCE, a substitute for creation 95.
CHANGE in nature 255.
CHILDREN, formation of judgment in 35, 36; dawn of consciousness of 166; intelligent perception in 170.
CHURCH, in relation to tradition 193.
COGNITION, its relation to feeling 65; necessary 91; congruity of 198.
COMMON SENSE and philosophy 111, 114; Reid's view of 208.
COMPREHENSION applied to judgment 18, 20.
CONCEPTIONS, isolated 38; symbols of objects 189; general 324.
CONCEPTUALISM, doctrine of 336; error of 337.
CONDITION, first 172.

CONSCIOUSNESS supposes a judgment 29; Spencer's opinion 33; Bain's postulate on 70; contingent 81; opinions on 81; material of 181; Mr. Huxley's opinion 280; degrees of 323; nominalists on 337; definitions of 340; where commencing 342; opinions on 343; scholastic doctrine of 351; direct and reflex 353; in acts of volition 356; objects of 359.
CONSERVATION to be admitted 95.
CONSISTENCY, principle of 70; of language 72, 73; a secondary test of truth 196; its want of value alone 199; its real dignity 213.
CONTINGENCY, definition of 56; its connection with physical certitude 93.
CONTRADICTION, principle of 65, 173; Mill's declaration of 73, 76.
CONTRADICTORIES, Hegelian use of 179.
COPULA, its value 20; equivalently in all speech 23.
CREATION to be admitted 95.
CREDULITY opposed to providence 106.
DELUSION, due to nervous derangement 237.
DIVINELY INFUSED IDEAS, a criterion of certitude 194.
DOUBT, definition of 44; Mill's definition of 44; negative and positive 44; etymologically considered 45; expelled 60; in physical certitude 99; methodic 148; Cartesian 151; a preliminary to certitude 152; St. Augustine on 160.
DUALISM, definition of 268; opposed to monism 299.
DUMBNESS of sceptics 139.
EGO, Mill's theory 176; man conscious of 352.
EMOTION, its effect on certitude 63, 67.
EMPIRICISM on association 74; on consciousness 81; confusion of 84; arguments of 86; opposed to scholasticism 113; result of 207, 256; in relation to memory 372.
ERROR, nature of 7; as distinct from ignorance 43, 233; supplementary causes of 237; due to want of thought 244.
EUCLID compared with philosophy 115.
EVIDENCE, in opposition to experiences 90; Mill's assertion 91; Reid's view of 208—210; the criterion *secundum quod fit judicium* 220; definition of 221; necessity of objective 225; objections against 225; mediate and immediate 227; when safe 229; necessitating 229; insufficient 241; internal 383.

EVOLUTION opposed to idealism 282.
EXISTENCE implicitly known 169; Mr. Spencer's theory of 182; to be proved 275.
EXPECTATION, Bain's postulate on 71.
EXPERIENCE, power to change truth, 78, 84; due to instinct 123; among animals 194; its value 223.
EXTENSION applied to judgment 18, 20; fallacy concerning 177.
EXTERNALITY, how reached by the mind 296.
FACT, primary 89, 169; establishment of 274; conscious 360.
FACULTY, each has its own excitant 222; finite 285.
FAITH, a criterion of certitude 191; the first act of intelligence 193; natural and supernatural 391; not a matter of intellect 393.
FEELING, its relation to cognition 65.
FORGERIES in history 387.
FREE WILL, abuse of 246.
GHOST STORIES, explanation of 239.
GOD and variability 93; supporting existence 94; Descartes' doctrine on 157; His Omniscience 222; alone *a se* 253; knowledge denied to 285; Berkeley's theory 294; Ferrier's opinion 295; revelation of 391.
GRACE, its action on the soul 392.
HABIT, its influence on the mind 124; productive of error 241, 245.
HALLUCINATIONS, opinions on 240.
HEREDITY among animals 194.
HISTORY, sacred and profane 383; ancient and modern 387; forgeries in 387; providence in 388.
IDEALISM, refuted 260, 278, 284, 312; protest of 269; difficulty of 273; self-contradicted 275; summary of 287; Berkeley's theory 294; idea of self 352; defined 303; dogmatism of 307; cosmothetic 310; in relation to memory 372.
IDEAS, definition of 5; meaning power of 5; Cousin's definition 6; Aristotle's definition 7; Spencer's ultimate 11; value of 105; not isolated 127; divinely infused 194; clear and distinct 195; innate inadmissible 204; not bounded by sensations 302; their position in space 307; their nature 307; intuitive 312; definitions of 312; universal 314; abstract 317, 328; individualizing of 329; physically one thing 318; never vague 321; isolation of 323.

IGNORANCE differing from error 8, 233; its divisions 43; extent of 234, 245; infinite 246; realized 394.
ILLUSIONS, causes of 237; Kantian doctrine on 362.
IMAGINATION, its powers 200; phenomenon of 231.
IMPENETRABILITY, Mr. Huxley's opinion of 286.
INCONCEIVABILITY, a criterion of truth 200.
INERTIA in intellect 205.
INFERENCES at fault 240.
INFINITE, THE, Hamilton's theory on 48.
INSIGHT, its connection with assent 53; into terms and their connection 80.
INSTINCT opposed to reason 120; experience due to 123; blind 188; of animals 225.
INTENSION, *see* Comprehension.
INTENTIONS, first and second 313.
INTELLECT, acts of 14; definition of 205; illuminates its object 218; how fallible 234, 245; distinguished from will 246; its action 235; its relation to consciousness 358.
INTELLIGENCE, its commencement 167; not self-explanatory 189; power of 205; of animals contrasted with human 225; unconscious 342.
INTENSITY OF CERTITUDE 59; denied 65; confused with size of object 67.
INTUITION, confusion respecting 310; definition of 327.
JUDGMENT, definition of 14, 22; how differing from apprehension 15; the crowning act of intelligence 17; theories on 18; its matter and form 21; characteristic of clear judgment 25; opinions on 28 seq.; of children 35; its identity with apprehension 38; controversy on nature of 51 seq.; not mere association 19, 74 seq.; apparent absence of motives in 125; sceptical suspension of 139; first 168; criterion of 218; false 234; erroneous, cause of 238; culpable 247; about external world 267.
KNOWLEDGE, its correspondence with object 2; symbolic correspondence of 3; partial, its value 4; cognitive reaction 9; Lange's opinion 11; mental assimilation 12; Spencer's opinion 33; differing from belief 47; as an equation 61; confined to consciousness 82; validity of 109; natural 111; interrelated 169; of self 171; de-

pendent on tradition 194; course of 195; process of 209, 217; God's and men's 222; growth of 234; transmitted 276; nature of 284; never absolute 285; mediate and immediate 310.
LANGUAGE useless to the idealist 111.
LOGIC, its relation to science 110; practical opposed to scholastic 121.
MAJORITY, consent of 194.
MAN, definitions of 7; compared with animals 27.
MATERIALISM compatible with idealism 293.
MATTER, solid, existence of 261.
MEMORY, Bain's postulate on 71; employed by idealists 308; habitual and actual 366; veracity of 368; limited 369; use of its admission 372; a peculiar faculty 373.
METAPHYSICAL CERTITUDE 55; contrasted with physical 56, 66; prejudice against 69; Bain's postulates 72; Mill's denial of 76; bound to necessity 89.
METAPHYSICAL PRINCIPLES, Descartes' doctrine on 157.
MIND, its method of working 125; empiricist destruction of 257; reaction of 285.
MIRACLES, objection to physical certitude 101.
MODES in substance 251.
MONISM, definition of 299.
MORAL CERTITUDE 57; its dependence on metaphysical 65.
MOTIVE, its definition 53; considered objectively 53; found in clear recollection 53; metaphysical 55; physical 56; expels doubt 60; of certitude always producible 122; apparent absence of 126.
MYSTICISM, supernatural and natural 129; a form of philosophy 161.
NATURE, physical 93; uniformity of 94; changes in 254; finite and infinite 314.
NECESSARY TRUTH, Huxley's interpretation of 69; question stated 73; of mathematical axioms 78; of our sensations 81; not facts of consciousness 82; answers to empiricists 83.
NECESSITY, metaphysical 55; physical 56; moral 57; law of necessity not reality 73; Hume's theory 88; and metaphysical truth 89; contingent 93; of revelation 395.

NEGATION, not the first act of thought 162 ; its intensity 61.
NEGATIVE, as a nonentity 60.
NERVES, stimulus 263 ; their action 238.
NOMINALISM, doctrine of 333 ; refuted 334.
OBJECT, how united with subject 217 ; material 230 ; immediate 311.
OPINION, definition of 45 ; not belief 48.
OPPOSITES in languages 180.
OTHERNESS of bodies 286; Mill's denial of 291 ; Huxley's denial of 299.
PERCEPTION, Spencer's definition 34 ; prior to language 36 ; requiring apperception 169 ; of self 170 ; distinct from sensation 265 ; sensitive 266 ; intellectual 267.
PERIODICITIES, uniformities of nature 97.
PERMANENCE of substance 253.
PERSEITY of substance 251 ; meaning of 253.
PERSONALITY, unknowable 11, 171 ; recognized 169.
PHENOMENALISM in relation to certitude 256.
PHILOSOPHY, its office in regard to certitude 58 ; its place in intellectual life 108 ; its definition 109 ; study of 110 ; position of scholastic 113 ; maxims 117 ; *versus* natural certitude 119 ; not co-extensive with practical discovery 124.
PHYSICAL CERTITUDE not mere possibility 56 ; bound up with metaphysical 65 ; Aristotle's contingent Being 92 ; answers to opponents of 102.
PHYSICAL SCIENCE, its basis 106.
PHYSICISTS elastic use of terms 124.
POPULAR ERRORS concerning Cartesianism 156.
POSSIBILITIES not physical certitude 56 ; have a real foundation 227.
PREDICATE relation to subject 18—21.
PRESENT, of an idealist 308.
PRIMARIES, the three 174.
PRINCIPLE, the first 173.
PROBABILITY, treatment of in mathematics 46 ; curious example of 49 ; used for moral certitude 58 ; used for certitude 122 ; in moral and intellectual matters 138 ; in place of truth 147 ; use of 229 ; its part in history 381 ; of revelation 393.

PROVIDENCE a factor in the world's physical course 102; in history 388.
PYRRHONISTS, doctrines of 135; Paschal's opinion of 212.
QUALITIES, primary and secondary 265, 282, 293.
RATIOCINATION, definition of 14.
REALISM, transfigured, theory of 3; definition of 268; unanimity with idealism 269; moderate 273, 278; summary of 287; a difficulty concerning 304; exaggerated 332.
REALITY, Mr. Hodgson's definition 10; Fichte's opinion 12; no objective 72; in matter, Mill's theory 176; of accidents 252; of substance 254; proof of external 281; the Unknowable 298; senses of 313; in universals 328; n scientific generalizations 338.
REASON, and instinct 120; its capabilities 133; an intrusted talent 143; *versus* instinct 260; its relation to faith 393—395.
RECOLLECTIONS of childhood 168.
REFLECTION the cause of scepticism 120; corrects judgment 126; our power of 357.
RELATION, ideas of 33; between objects 283; in knowledge 169, 180.
RELATIVIST, his position 3.
RESISTANCE, the only external activity 3.
REVELATION, its rare occurrence 129; belief in 391; probability of 393; a necessity 395.
SCEPTIC, description of 134, 135.
SCEPTICISM, cardinal error of 5; due to reason 120; account of by Sextus Empiricus 135; dogmatic and non-dogmatic 136; a sin against intelligence 142.
SCHOLASTICS, definition of truth 9; on apprehension and judgment 38; on metaphysical necessities 55; on doubt 44; position of their philosophy 113; definition of intellect 205; on the generation of knowledge 217; on fallible intellect 234; on substance 250; on universals 318.
SCIENCE, definition of 109.
SCIENCES, on classification of 116.
SCRIPTURE, appeal to isolated passages of 146.
SELF, certainty of 154; perception of 170; consciousness of 176; definition of 351; animal appreciation of 353.

SELF-CONCEIT of philosophers 162.
SELF-CONSCIOUSNESS defined 341.
SENSATION prior to language 36; confounded with intellectual perception 77; truth of 81; Lewes' theory of 189; Mill's theory of 190; distinct from perception 265; definition of 266; never false 288; permanent possibilities of 289; Helmholtz's theory 326; without consciousness 343.
SENSES, their trustworthiness 2, 288; verification by 190; validity of 259; division of 262; value of 276; gradual training of 298.
SENSIBLES, division of 264.
SIGNUM QUO, A QUO, EX QUO 5.
SOLIPSISM, difficulty of 293.
SOUL, definitions of 7; proof of its spirituality 111.
SPACE, extension in 286.
SPECIES, Aristotle's definition 54; intued 310.
SUBJECT relation to predicate 18—21; how united with object 217.
SUBJECTIVITY valueless alone as criterion of truth 224.
SUBSTANCE, scholastic notion of 250; efficiently active 255.
SUFFICIENT REASON, principle of 175.
SUSPENSION OF JUDGMENT, employed by sceptics 139.
SUSPICION, description of 45.
SYLLOGISM useful for verification 126.
SYSTEMS, definition of 150.
TASTE, referable to subjective conditions 231.
TESTIMONY, belief in 377; reasons for believing 378; of a single witness 381, 387; cautions respecting 381.
THEISM, necessary 95; itself philosophic 98.
THEORY and practice 123.
THOUGHT, essentially inaccurate 132; unconscious 133; first act of 162; not to be isolated 167; known only by speech 192; mystery in process of 274; and thing 305; objective validity of 309; without consciousness 343.
TIME, alters nothing 94.
TOUCH, division of 262.
TRADITION, a criterion of certitude 191; objection to 192.
TRUTH, division of 1; ordinary view of 2; schoolmen's definition 2; opinions on scholastic definition of 9; theological

importance of definition 12; where found completely 14—16; metaphysical 55; as an equation 62; necessary truth, meaning of 69; Bain and Huxley on 70; geometric 77; not sacred 83; confusion respecting 84; mathematical distinguished from physical 86; self-evidence a motive of belief 122; Descartes' doctrine on 156; self-evident 174; never inconsistent 196; ontological 221; present even in falsehood 245; intuition of 310.

UNIFORMITY OF NATURE 94; Dr. Maudsley's opinion 104.

UNIVERSALS, explanation of 314; how formed 316; sensist theory 319; where found 321; difficulty against 322.

VARIATION, concomitant 4; in degrees of certitude 63.

VERIFICATION by senses 189.

WILL, its action calculated 57; commanding assent 64; contrasted with "association" 74; its action on the intellect 235—242.

WORDS no test of universals 321—334.

---

## LIST OF AUTHORS REFERRED TO.

ANSELM (St.), knowledge mental assimilation 12.

ANTISTHENES on judgment 19.

AQUINAS (St. Thomas), definition of soul 7; nature of the mind 9; knowledge mental assimilation 12; on judgment 17; value of copula 20; on clearness of judgment 24; contrast between judgment and animal perception 26; on sensitive faculty 36; apprehension *aliquale judicium* 38; on knowledge as an equation 61; on the generation of knowledge 217; on sensibles 264; all thoughts must be consciously referred to self 348; on acts of the mind 355.

ARISTOTLE, definition of idea 7; definition of soul 7; value of copula 20; on species 54; on Being 92; on substance 254; on sensibles 264.

AUGUSTINE (St.) on God's Providence 102; on doubt 160.

BACON (Lord) on proposition 33.

BAIN (Mr.) on the mark of judgment 26; on the principle of consistency 70; on spontaneous beliefs 75; on mathematical axioms 79; on truth 84; denial of efficient causality 94, 96; on blind tendency 106; on Hume's consistency 144; on principle of Contradiction 173; on isolated events 175; on substance 253; on the perception of matter 271; 274 *n.*; on idealism 282; on ordinary realism 287; on existence beyond consciousness 306; on consciousness 345; on belief in testimony 378;

BALFOUR on Mr. L. Stephen 85; on Mill's idealism 290.

BALMEZ on certainty 110; on errors 118; experience due to natural instinct 123.

BAYLE on instinct and reason 120; faith against reason 191.

BERGMANN, definition of judgment 26.

BERKELEY (Bp.) on ideas of sense 270; theory of idealism 294; on nominalism 334.

BOSSUET on Cartesianism 159.

BRADLEY (F. H.), definition of judgment 19.

BROWN (Dr.) on apprehension and judgment 30; on Providence 105; on belief 210; on consciousness 348; on memory 373.

BROWNE (Mr. Borden) conception of reality 9.

BUCHNER on Hegelianism 180.

BUCKLE on Descartes 158; on history 389.

BUFFIER on ideas 312.

CAIRD (Prof.) on validity of a truth 198; on Berkeleyism 295; on idealism 304.

CARLYLE (T.), use of certain plurals 21.

CLIFFORD (Prof.) on the mark of judgment 26; on our laws of geometry 80; on possibilities 95; on cause 259; on belief in an outer world 261; on other consciousnesses 292, 299, 305.

COMTE (M.) on consciousness 82; on the absolute 86.

CONDER (Mr.) on consciousness 360.

CONGREGATION OF THE INDEX on reason 133.

COUSIN on idea 6; criticism on Locke 40; judgment, the primitive elements of thought 40; on scepticism 161; on the genesis of error 236; on space 332; on consciousness 351.

DE BONALD, thought known only by speech 192.

DE LAMENNAIS on traditionalism 192; on Diderot 214.

DE MORGAN (Prof.) on belief in concepts 39; on certainty 101;

DESCARTES on separation of assent from its motives 51 ; on influence of the demon 102 ; doubt of self-evident truths 113; system of 150; on truth 156 ; real position of 159; as a physicist 159 ; on personality 170 ; on the genesis of error 235; on action of the senses 267; definition of ideas 312.
DÖLLINGER (Dr.) on forgeries in history 387.
DRAPER (Mr.), credibility of 383.
EMPEDOCLES, like known by like 8.
EMERSON, on want of coherence 213.
FERRIER on the use of philosophy 112 ; on Hume's philosophy 144 ; on intelligence in children 166 ; on being and not being 177 ; on the perception of matter 295 ; consciousness distinct from sensation 344.
FICHTE, reality a dream 12 ; on idealism 262.
FINDLATER (Mr.) on universality of copula 23.
FISCHER (Kuno) on causality 259 $n.$; on nature of knowledge 284 $n.$; on the Ego 363.
GOETHE on union of mental conceptions 127 ; on transcendentalists 180.
GREEN (Prof.) on Hume's consistency 145; our process of learning 212 ; on cause and substance 259 ; on Berkeleyism 294.
GROTE on the canon of evidence 232.
GURNEY (Mr.), objection to universality of copula 22.
GUTBERLET (Dr.), certitudes are equations 61.
HAMILTON (Sir Wm.) on apprehension and judgment 29 ; on belief and knowledge 47 ; on proof of mendacity of consciousness 142 ; on Hume's consistency 144 ; on the error of Descartes 149 ; on the foundation of knowledge 211 ; on error 236, 242 ; on sensation and perception 265 ; on primary and secondary qualities 265, 293 ; on idea and perception 311 ; on consciousness 341, 348, 359 ; on memory 374.
HARRISON (Mr. F.) on the correspondence of conceptions with reality 3.
HARTMANN on the Great Unconscious 363.
HEGEL on philosophy 112 ; on contradictories 179 ; on the mind 259 ; on abstract ideas 328.
HELMHOLTZ on sensations 326.

HERBERT (Lord) on traditionalism 192.
HERODOTUS, instance of inference by rapid association 126.
HOBBES, definition of judgment 18 ; on Nominalism 333, 334.
HODGSON (Mr. S.) definitions of truth and reality 10.
HOLLAND (Sir H.) on memory 373.
HUET (Bp.) on non-dogmatic scepticism 145.
HUME on idea 5 *n.*; on apprehension and judgment 28 ; on inferences from experience 86 ; on necessity 88 ; on fact 89 ; on Cartesian doubt 113 ; denies existence of sceptics 135 ; on probability 139 *n.*; his consistency 144 ; on Descartes 152, 155 ; on a future state 202 *n.*; on idealism 270 ; on universals 335.
HUTCHESON on consciousness 343.
HUXLEY (Prof.) on necessary truth 69 : no objective reality 72 ; self-contradiction 78 ; on consciousness 81, 82 ; on truth 84 ; on necessity 88 ; on denial of insight 90 ; on uniformity of nature 99 ; on substance 253 ; on realism 272 ; on idealism 280, 293 ; on impenetrability 286 ; on materialism 293 ; on otherness 299 ; on ideas 302 ; on consciousness of sensation 350 ; on memory 368.
JEVONS (Mr.), logical foundations of languages 24 ; on Mill's inconsistencies 106.
JOSEPHUS on history 384.
JOUFFROY on instinct and reason 120.
KANT, doctrine on judgments 32 ; on organic processes 342 ; on consciousness 350, 362 ; on illusions of reason 362.
LAMB (C.) on the sceptic 134.
LANGE on value of our knowledge 11.
LECKY (Mr.) on error 237.
LEIBNITZ on empiricism 205 ; on unconscious ideas in plants, &c. 344.
LEPIDI, criticism of judgment 218.
LEWES, perception in animals 27 ; on grouping ideas 32, 35 ; on certitude 62 ; on efficient causality 94 : on blind instinct to believe 189 ; on ideas 302 ; on consciousness 345.
LIBERATORE (Fr.) *signum a quo* 5.
LITTRÉ (M.) on history 389.
LOCKE on judgment without perceivable proof 39 ; on simple ideas 223 *n.*; on consciousness 348.

LOTZE on sensation 265.
LUYS on evolution of thought 35.
LYELL (Sir C.) on omissions in history 386.
MAHAFFY (Mr.) on historic belief 380.
MANILIUS on reason 235.
MALEBRANCHE on ontologism 195.
MANSEL, definition of truth 10; on apprehension and judgment 29; on psychological judgment 38; on sufficient reason 175; on consistency 196; on Hamilton's distinction between conception and belief 211; on objective criterion of truth 219 *n.*; on error 237.
MARTINEAU (Dr.) on omissions in history 386.
MAUDSLEY (Dr.) on consciousness 82; on the theory of association 103; on arts of consciousness 126; how biassed 300; on general ideas 331; on consciousness 346; on memory 374.
MAURUS (Silvester) on the Blessed Trinity 13; on the generation of knowledge 218.
M'COSH (Mr.) on cognition 36, 37; on foundations of knowledge 172; on ignorance and error 237: on consciousness 349.
MILL (James), definition of judgment 18; on process of perception 35; on consciousness 350; on human testimony 377.
MILL (John S.), definition of judgment 18; on proper names 20; belief the characteristic of judgment 25, 26; on propositions 33; on doubt 44; on association 74; negation of certitude 76; on mathematical axioms 79; on truth 83; his estimate of logic 87; on evidence 91; on uniformity of nature 98; inconsistencies of 106; on language 111; on probabilities 139 *n.*; on scepticism 141, denies existence of sceptics 143; on axiomatic truths 155; on principle of contradiction 173, 179; on cause and effect 175; on the Ego 176; on sensation 190; on substance and accidents 252; on efficient causality 255; on the senses 271; permanent possibilities of sensation 289; on memory 308; on universals 322; on the isolation of ideas 323; on general concepts 323, 336; on consciousness 349.

MORELL on sensation and perception 36.
MORLEY (Mr.) on Mill as a teacher 107.
MOZLEY (Mr. T.) on Tractarianism 131.
MÜLLER (Max), forms of grammar common to all nations 23 ; on doubt 45.
NEWMAN (Card.) on assent 46 ; on proportion in assents 61 ; opposed to Descartes 150 ; on acts and faculties 172 ; on consciousness 347 ; on Greek historians 385.
NICHOLAS OF CUSA (Card.) on impotence of reason 145.
PAGET (Sir James) on the complexity of human acts 364.
PALMIERI (Fr.) on direct consciousness 354.
PASCHAL on pyrrhonists and dogmatists 212.
PASTEUR (M.) on M. Littré 390.
PATTISON (Mark) on Tractarianism 132.
PAUL (St.) on faith 395.
PHILETAS and dogmatic scepticism 131.
POLLOCK (Mr.), denial of efficient causality 94 ; on the value of ideas 105 ; on systems 150.
PORTER (Dr.) judgments of children 35.
READ (Mr. C.) on doubt of actual cognitions 94.
REID on apprehension and judgment 29 ; on the fixed course of nature 105 ; on Descartes' system 152 ; on criterion of truth 208 ; on cause 259 ; on sensation 266 ; on consciousness 343 ; on memory 372.
RENAN (M.) his mental diagnosis 130.
RIBOT (M.) on value of philosophy 112 ; on memory 376.
RICHTER on intelligence in children 166.
ROSMINI, approval of Kant 32.
ROUSSEAU on ignorance and error 233 *n*.
SALISBURY (John of), description of sceptics 136; on knowledge of truth 191.
SAYCE (Mr.) on universality of copula 24.
SCHOPENHAUER on Hegel's philosophy 180 ; on the attributes of bodies 282.
SEXTUS Empiricus on scepticism 135.
SIDGWICK (Mr. A.) on theory and practice 121.
SPENCER (Mr. H.) transfigured realism 3 ; ultimate ideas 11 on the formation of an idea 35 ; on consciousness 81, 327 ; on proof of mendacity of conciousness 142 ; on

classification of sciences 116; on absence of motives in judgment 127; on isolation of thought 167; on certainty of self 171; on development of intelligence 181; on philosophy of certitude 185; on congruity in cognitions 197; his inconsistencies 200; universal postulate 204; on substance 253; on idealism 282; on the Unknowable Reality 298; theory of memory 375.
SPINOZA on idea 5; on true ideas 196; on consciousness 350.
STEPHEN (Leslie) on truth 83; his only dogmatism 85.
STEWART (Dugald) his law on concepts 39.
SUAREZ, knowledge mental assimilation 12; on judgment 38; on the generation of knowledge 218; on error 246.
SULLY (Mr.), objection to universality of copula 22; judgments accompanied by belief 26; on recognition in animals 27; on judgments in children 36; on judgments 127; on hallucinations 241; on introspection 357.
TAINE (M.) on illusions 240; on external objects 292.
TAIT (Prof.), proof of external reality 281.
TAYLOR (Mr.) on omissions in history 390.
TYNDALL (Prof.) on imagination 201; on human ignorance 394.
UEBERWEG, definition of judgment 26.
VEITCH (Prof.) on modes of thought 242 *n*.
VINCENT OF BEAUVAIS on Universals 330.
WALLACE (Mr.), unity the test of truth 198.
WARD (Dr.), refutation of empiricists 85; on analysis of grounds of truth 121.
WHEWELL on Descartes' physical science 159.
WHITMAN (Walt.) want of coherence 215.
WUNDT on judgment 37; opposed to Descartes 150.
WYLDE (Mr.) on connection of mind with mind 230.
ZIGLIARA (Card.), definition of judgment 22; on assent in judgment 50.

www.ingramcontent.com/pod-product-compliance
Lightning Source LLC
Chambersburg PA
CBHW030545300426
44111CB00009B/865